T0099977

Earth's Emergency Room

OTHER BOOKS BY LOWELL E. BAIER

Inside the Equal Access to Justice Act: Environmental Litigation and the Crippling Battle over America's Lands, Endangered Species, and Critical Habitats (2016)

Saving Species on Private Lands: Unlocking Incentives to Conserve Wildlife and Their Habitats (2020)

Federalism, Preemption, and the Nationalization of American Wildlife Management: The Dynamic Balance between State and Federal Authority (2022)

The Codex of the Endangered Species Act, Volume I: The First Fifty Years (2023)

The Codex of the Endangered Species Act, Volume II: The Next Fifty Years (2023)

EARTH'S EMERGENCY ROOM

Saving Species as the Planet and Politics Get Hotter

LOWELL E. BAIER

ROWMAN & LITTLEFIELD
Lanham • Boulder • New York • London

Published by Rowman & Littlefield
An imprint of The Rowman & Littlefield Publishing Group, Inc.
4501 Forbes Boulevard, Suite 200, Lanham, Maryland 20706
www.rowman.com

86-90 Paul Street, London EC2A 4NE

British Library Cataloguing in Publication Information Available

Library of Congress Cataloging-in-Publication Data
Names: Baier, Lowell E., author.
Title: Earth's emergency room : saving species as the planet and politics get hotter / Lowell E.
   Baier.
Description: Lanham : Rowman & Littlefield, [2024] | Includes bibliographical references and
   index. | Summary: "Drawing on his extensive experience as a prominent environmental lawyer
   and activist, Lowell Baier captures the colorful and important history of the Endangered Species
   Act and argues that it can be a powerful tool to ameliorate the biodiversity crisis while still
   respecting landowners, states, and industries"— Provided by publisher.
Identifiers: LCCN 2023053792 (print) | LCCN 2023053793 (ebook) | ISBN 9781538194133
   (cloth) | ISBN 9781538194140 (epub)
Subjects: LCSH: United States. Endangered Species Act of 1973—History. | Biodiversity
   conservation—United States. | Biodiversity conservation—Government policy—United States. |
   Endangered species—United States. | Endangered species—Government policy—United States.
   | Endangered species—Conservation—Government policy—United States.
Classification: LCC QH76 .B34 2024 (print) | LCC QH76 (ebook) | DDC 333.720973—dc23/
   eng/20231222
LC record available at https://lccn.loc.gov/2023053792
LC ebook record available at https://lccn.loc.gov/202305379

♾™ The paper used in this publication meets the minimum requirements of American National
Standard for Information Sciences—Permanence of Paper for Printed Library Materials, ANSI/
NISO Z39.48-1992.

Contents

List of Illustrations

FOREWORD: A SPIRITUAL CALLING TO CARE

Fifty students gathered at the Harvard Divinity School for a class called "A Wild Promise." Students were from across the campus: from the Kennedy School of Government, the School of Design, the Medical School, the School of Education, and the Business School. Younger students from Harvard College were also represented. And there were those from the Advanced Leadership Initiative, many of them industry titans who had made a fortune for themselves and now wanted to return to a larger world of ideas and disciplines and give something back to a society that had given so much to them. They enrolled in this class to celebrate the fiftieth anniversary of the Endangered Species Act and why we should care about threatened and endangered species.

Who better to introduce this visionary piece of bipartisan legislation signed into law on December 28, 1973, than the distinguished author, legal and environmental historian Lowell Baier.

Mr. Baier appeared on a screen via Zoom. He had on a tweed jacket and a button-down shirt with a bow tie, with a wall of books behind him. There was a candle lit on his desk. For thirty minutes, he spoke about the history of the Endangered Species Act and its co-signatories (including his duck hunting partner, Representative John Dingell from Michigan) and how respect ruled in Congress and, though disagreements can occur, pragmatic vision can be the result of compromises made in the name of the greater good. Support for the Endangered Species Act was an almost unanimous vote in the House of Representatives, with full support found in the United States Senate—difficult to imagine in today's Congress. Baier spoke about how the Endangered Species Act began as something aspirational to care and protect the plants and animals, vertebrates and invertebrates among us, that are vulnerable. But it became something tangible, real, and visionary, a moral imperative, to refuse extinction and adopt the soaring belief that every living organism should be preserved. The Endangered Species Act ensures this right through legal protections on private and federal lands.

"Extinction is worse than death," Baier said.

Then he spoke of how "an estimated 291 species have been saved from extinction by the Endangered Species Act, directly." He told us at present there are "1,686 species in the United States that are listed as either threatened or endangered." He filled out his facts by explaining in "the last ten years, 258 species have been listed, while just 55 species were delisted," the bald eagle and American alligator among them. It is a law that has been 99 percent effective as a stay against extinction.

But the list of threatened species is growing longer every year. And in 2023, the U.S. Fish & Wildlife Service delisted twenty-one species due to extinction. The list included the Little Mariana fruit bat in Guam, two species of fish (the San Marcos gambusia in Texas and Scioto madtom in Ohio), and ten species of birds (including the Bachman's warbler, a small yellow and black songbird, once found in Florida and South Carolina; the bridled white-eye, a green tropical forest bird from Guam; and eight honeycreeper species in Hawaii), along with eight species of mussels. Gone forever.

After his electrifying lecture, Lowell Baier paused, as if he were looking past us, wondering whether he dared to say what he really wanted to say.

"Why is the Divinity School concerned about the loss of biodiversity, about climate collapse and the Earth crisis?" he asked. "I've been thinking about this." He paused again. "You are students studying at the Divinity School. So, here is my challenge to you." All eyes were on this man whose career path included being a real estate developer, an attorney, a fierce advocate for wildness, a self-described centrist, and a consultant to both Republican and Democratic presidents on issues of wildlife conservation.

"I am in my ninth decade as one who has devoted the latter part of my life to understanding and defending the Endangered Species Act. But I am struggling . . . I am struggling with what we need to do *now* to protect the act itself from those who care more about the supremacy of our own species at the expense of all other species; lawmakers and lobbyists are more attuned to capitalism than conservation."

He focused directly on the students in the room. "Here is what I can tell you: Politics is not enough. Science is not enough. Neither inspires nor creates a heartfelt conversion to care about plants and animals who are facing extinction. What is missing is the spiritual conversation. What is missing is a spiritual language that can move us to care."

I watched as the students leaned forward to listen more closely.

"Will you accept my challenge and find that spiritual language, dare to have this critical conversation about the sacred nature of Earth and how the health of our planet is the health of all life, not just our own? Speak to our hearts—this is what is missing, and we need it."

After Lowell Baier delivered his inspiring and empowering lecture, students lined up to the microphone to ask him questions—the spirited conversation between the students and Mr. Baier went on for an hour until it was time to leave.

The agenda was set for the class. "What would our wild promise be and to whom?" Each student picked an endangered species to study, contemplate its niche within its ecosystem, the causes of its demise, and how restoration of its population might occur. Students were encouraged to be creative with the final project, creating a poem, program, film, article, or curriculum around it, using whatever analytical or creative means available to them with the gifts that were theirs.

After watching films on endangered species such as *Albatross*, directed by Chris Jordan; *The Path of the Panther*, by Eric Bendick; *All That Breathes*, by Shaunak Sen; and *Fathom*, by Drew Xanthopoulus (alongside the film *Generation Wealth*, directed by Lauren Greenfield, which looks at the various manifestations of greed at the expense of the planet), students presented their projects. Every single student accepted Lowell Baier's challenge to engage with the spiritual components of the Endangered Species Act. They spoke from their hearts, told stories, and it was deeply personal.

Here are four examples.

Miriam Israel took the Hebrew phrase "the whole world is a very narrow bridge, and the challenge is to not be afraid," spoken by Rabbi Nachman of Braslav, a Hasidic scholar, to better understand the narrow constraints of what it means to be human in the face of wild beauty that is threatened. The species she chose to study was the African gray crowned crane, which she writes "can often be found perched on the top of acacia trees, balancing carefully on the narrowest of branches" of the few trees that have survived the onslaught of development. "The sight of these large birds, sitting so gracefully on these tiny branches, reminded me instantly of the idea of a 'narrow bridge,' and as the events in Gaza have unfolded over the past few months, I have been unable to get the two images out of my head."

Rey Chin created a project called "Imaginal Migrant"—in her words, "a visual art piece that integrates the spatial and temporal dynamics of the eastern and western migratory monarch populations into a single artifact. It is a hand-cut and crafted, three-dimensional monarch butterfly replica made of layered black and orange paper." By integrating scientific data points with artistic expression, she created a model of the butterflies whereby their migratory paths become the black veins mapped on the orange wings. She used the evocative word "homegoing" as an interpretation and integration of how one species can captivate the imagination of another species through devotion. A monarch butterfly's "homegoing" becomes a "homecoming" for humans who care enough about the monarch's well-being to plant milkweed along their migratory routes.

Jeremy Barber spoke to the plight of the mountain lion becoming locally extinct in wildlands adjacent to Los Angeles due to inbreeding as a result of highways and freeways separating the feline populations. He then made a comparison of the lion's plight with the isolation of marginalized human communities and how they are also cut off from one another—"Freeways are not free," he writes, giving the shadowed history of why and how our "concrete arteries have destroyed communities," both human and wild, thwarting the necessary movement and migration in order for life to thrive. He called for an ethical stance in which all communities can flourish in the name of empathy and compassion.

Finally, Alex Kinsella, an attorney who has worked with the Diné in the Four Corners region of the American Southwest, took the Endangered Species Act and reimagined it with spiritual language and practices conducive toward healing the divide between human and other life forms on Earth. She writes, "In my attempt to rewrite parts of the Endangered Species Act, I draw on Donna Haraway's imaginative prose (*Staying with the Troubles*) . . . to center the relationship between humans and non-human beings. Building from her assertion that we must bring the dead into active presence before we can recuperate, I placed collective mourning a primary purpose of the Act. I further draw from the structural approach of the Council of All Beings ('Council'), which is a deep-ecology series of re-earthing rituals to help humans deepen their empathic relationship with the natural world. The Council engages in three processes during their ritual gatherings: mourning, remembering, and speaking for other life-forms."

Throughout her project, Kinsella writes, "I asked myself: How can the law and the imagination weave together to rematriate human relationships with non-human beings? What perspectives, human and non-human, are missing from current relevant legislation? What can the Indigenous communities with distinct relationships to the natural world and its many beings teach the broader American public about balanced relationships between humans and non-human persons?"

These poignant examples are illustrative of the impact Lowell Baier had on our students to bring spiritual insights into the powerful and restorative history of the Endangered Species Act. He saw these students as spiritual practitioners working within "Earth's Emergency Room."

Baier writes in this impassioned book of critical care and collaborative actions on behalf of threatened and endangered species: "This is a call to action of the obligations to ourselves, our families, and society for making extinction, biodiversity, and the ESA a national priority again."

We can and we must. We are living in the Sixth Extinction. This is not just a political crisis or even an ecological crisis—it is a spiritual one. Who will we become when the elegant lives around us, from grizzly bears to white-bark pine to the rusty-patched bumblebee to elkhorn coral, seen and unseen, known and unknown, vanish under our watch? And if we do nothing? Imagine the vast loneliness that will reign over us.

A world without monarch butterflies is not a world I wish to contemplate.

On the last day of class, Lowell Baier met with the class once again. His wisdom and passion for the Endangered Species Act framed our course, "A Wild Promise." This time, he was wearing a red cardigan for the holiday season. When he appeared on the large screen smiling, he was greeted by the fifty students and a standing ovation. Moved by their gesture, he urged them to make their own passionate engagements toward justice, even in the midst of the Israel-Hamas war, and recognize the intersection between a peaceful world and peaceful planet. He said, "I exhort you to support a second 'Green Revolution' like the one we saw in the 1960s and 70s," only this time, he said, "it will incorporate a holistic approach to an ethical stance toward all life—social justice is environmental justice for all." He concluded, "I will end by saying what Oliver Wendell Holmes once said: '*In our youth our hearts were touched with fire.*' May your hearts be touched by fire! The Endangered Species Act will continue through love. I wish you all well."

Nobody wanted to leave—and then one student named Elijah started to sing "Amazing Grace," and everyone formed a circle and sang.

"Together," Lowell Baier writes in *Earth's Emergency Room*, "we can address the biodiversity crisis, prevent extinctions, and begin to heal Earth of the wounds inflicted by our civilization."

Amen.

Terry Tempest Williams
Castle Valley, Utah
December 28, 2023

Introduction

Since my farm boy and Eagle Scout days in Indiana and my early childhood summers spent on a Montana ranch, I have cherished nature. Admittedly, after working my way through college and law school, I paved over many acres of urban land as a Washington, D.C., real estate developer. Like many people who need to earn a living, I've wrestled with the tension between human population growth and the conservation of our planet. An avid sportsman who spent many weeks afield, I respect nature's delicate balance. As my concern for this balance has increased exponentially over the years, I've become wholly committed to land stewardship and legislation that bridges the divide between people's needs and the planet's health. Now, in my ninth decade of life, Earth is experiencing an emergency that has evolved so slowly that most don't recognize it, brought about by habitat destruction and climate change, resulting in the rapid loss of many of the world's species.

Although I never abandoned my interest in conservation, I developed well over a million square feet in some fifty projects in the name of Baier Properties, Inc., over fifty years' time. I built and managed warehouses, shopping centers, office buildings, and related sites for gas stations and restaurants. I've lived through the traumas of commercial real estate development, through its highs and lows, from high risk and reward to the marketplace vanishing and leaving you with an empty building, a big mortgage, and the prospect of bankruptcy. Moreover, I've continuously lived through the never-ending dialogue and pressures to protect endangered species and the perils of those who constantly try to gut the laws protecting those species. That advocacy has become my driving passion.

As busy as my real estate career was, I became aligned with wildlife habitat research and field surveys, many of which I underwrote. Instead of a backpack and gun, I carried microscopes and analytical biological kits for examining skeletons, feces and droppings, and carcasses across backcountry places on the North American continent and central Asia. Instead of hunting companions, I was accompanied by wildlife biologists, botanists, and wildlife managers to evaluate habitats, animal populations, forage, and predators. We went to many places where I'd gone as a hunter, remote from people except nomadic summer livestock herders going into the high country where they could find pasture and shepherd domestic sheep and cattle. Local guides were always with us to find the wild animals.

These trips were educational experiences, as I learned how the locals denuded the land, killed off threatened and endangered wildlife to eat, and protected their herds from predators. I learned about wildlife's cycle of life, vegetation and habitat preferences, and migration patterns. Few conservationists and biologists have this diverse field experience across the world beyond their university education.

I've learned that over the past fifty years, more species have gone extinct than all of the species that have gone extinct since man began discovering them from fossil records centuries ago. Forty percent of America's animals now face extinction, as do 34 percent of plants. Forty-one percent of our *ecosystems* are facing collapse.[1] Birds—perhaps the most visible, most studied, and most conserved of all our species—illustrate the depth of the biodiversity crisis. Since 1970, we've lost 3 billion breeding adult birds in the United States and Canada across 529 bird species—25 percent of our bird population from twelve bird families alone.[2] The eastern monarch butterfly population has decreased by 80 percent and the western by 90 percent.[3] And between 2022 and 2023, beekeepers lost 48 percent of the managed honeybee population.[4] Moreover, 25 percent of wild bee species are at risk of extinction.[5]

Birds, butterflies, and bees; *so what?* you might ask. Species extinction isn't just an emotional loss to humans; it's also a threat. Twenty species provide 90 percent of the world's food, and just three—wheat, maize, and rice—provide more than half.[6] Birds, bees, and butterflies are the pollinators of our food, and half of pollination comes from bees alone.[7] Corn and wheat are pollinated by the wind, but all the rest are at risk, as pollination is threatened. Hence, the security of our food supply is at risk because our pollinators are dying.

It might surprise you to learn that 50 percent of all critical medicines are derived from plants and animals, including nine of the ten most prescribed in the United States. The Madagascar periwinkle produces chemotherapy drugs that treat Hodgkin's disease, acute lymphocytic leukemia, and several other cancers. Aspirin is derived from the herb meadowsweet. Malayan pit vipers and bloodsucking leeches produce venom anticoagulants. Horseshoe crab blood is used to detect contaminating bacteria in vaccines, injectables, and other sterile pharmaceutical equipment. And, of course, penicillin was derived from ingredients in blue cheese.[8]

With our food supply and critical medicines at risk from extinction of so many species, humankind is put in a vulnerable position. This idea might seem hard to believe since most of us don't experience daily deprivation—yet. But until the 2020s, few of us believed in the modern age that we could be vulnerable to global supply chain disruption and an atrophying manufacturing sector, making it difficult for us to get our car repaired or prescription filled.

"Biodiversity" is the totality of the natural world around us and the variety of all the different kinds of organisms—the plants, animals, insects, reptiles, crustaceans, amphibians, fish and aquatic species, and microorganisms that live on our planet, including man. All of these species live and work together in a diversity of ecosystems to maintain and support life on Earth, and all exist in a delicate balance.

Those who remain skeptical of Earth's looming emergency should examine the images from NASA satellites tracking changes to life on Earth. Woody Turner, program scientist for NASA's Biological Diversity Research Program, says, "We are really at a global biodiversity crisis, losing not only entire species but also seeing decreases in the number of plants and animals that are important for natural ecosystems."[9]

Now consider climate change, which—driven by human activities—exacerbates habitat loss (which takes ages to reverse) and consequently species loss (which is irreversible). As industry and greenhouse gas emissions (principally CO^2) have grown exponentially over the past 100 years, Earth's temperature has increased by 1 percent. That's more than it increased over the previous 6,000 years. As population increases and industrialization expands to keep up with this growth, CO^2 and other greenhouse gas emissions increase.

By 2050, Earth's temperature is projected to increase another 2.7 degrees Fahrenheit, and sea levels will rise ten to twelve inches—putting 10 million people in the United States at risk from coastal storms and flooding. Droughts will become more pronounced as fresh water dries up. Depending on which model is used for forecasting, Earth's temperature will rise another three to twelve degrees by 2100. As I write this book in the summer of 2023, global temperatures have shattered new record highs never seen on Earth, and they're going to keep climbing.[10]

What species can survive catastrophic fast-moving wildfires, 100- and 500-year floods, snowpack so deep that animals can no longer forage for food, extreme droughts, rising sea levels, and increasing temperatures, especially at higher elevations? "A warming planet endangers Americans and people around the world—risking food and water supplies, public health, and infrastructure and our national security," warns the Biden White House's 2022 National Security Strategy report.[11]

I am convinced that self-interested commerce, science, and technology will not on their own neutralize the combined effects of climate change and the biodiversity crisis. The challenge before us will require collective will and action—but the kind of will and action that respects people's right to earn a living and manage their private property.

We don't need to reinvent the wheel to achieve this outcome. The biodiversity crisis, accelerating because of climate change, can be mitigated by the Endangered Species Act of 1973 (ESA)—a bipartisan law that was passed by Congress and signed by President Richard Nixon when I was a young Washington lawyer supporting its creation. The stated purpose of the ESA is to protect Earth's species at risk of extinction and their habitats. The act is the emergency room of the biodiversity crisis. It gives legal protection to species threatened with extinction until they are rehabilitated. When the act is working as it was intended, species go into Earth's emergency room for a finite period of time and ultimately exit healthy enough to expand their population on their own. The web of life is thereby repaired, one species at a time.

But tragically, in today's political climate, the ESA doesn't function as well as it could. Congressional neglect, growing political polarization, and shortsighted special interest groups have weakened the once very effective act. Imagine your own community's once-vaunted emergency room now mired in political controversy and in a state of neglected disrepair.

Although I never became a graduate biologist, I learned much of their language. This knowledge aided me in understanding biodiversity and the role of the ESA in protecting it. On a pro bono basis, I've evaluated wildlife legislation and advised Congress, the White House, the Department of the Interior, and the Department of Agriculture. I became sufficiently well known in Washington as a centrist without a hidden agenda and was asked by President George H. W. Bush to prepare his administration's wildlife conservation agenda. Thereafter, each president—Republican and Democrat—has engaged my counsel on wildlife conservation legislation.

I have found myself in the middle of many dialogues and can appreciate the continuing conflict between capitalists and conservationists. Both have legitimate positions, but compromise can be reached only by checking egos and private agendas at the door and finding common ground, which is frequently not easy. To the detriment of the ESA and sensible conservation efforts, political extremists on both the right and the left (though mostly on the right) have become more influential. As we approach the 2024 elections, my aim is to impart renewed respect for the ESA and to convince you to join me in protecting it against future abuses and attacks.

Bipartisanship ruled the day in 1956, when Congressman Charles A. Halleck appointed me to be his page boy in the U.S. House of Representatives. Halleck took a liking to me because I was the first boy from Jasper County in his congressional district to attain the rank of Eagle Scout. I was sixteen, living on a hardscrabble grain and livestock farm in northern Indiana without any connection to the nation's capital or the political world.

My assignment was to shadow him throughout his day in the Congress and on social affairs after hours and on weekends as his chauffeur, briefcase carrier, notetaker, and messenger, as we had no BlackBerrys, iPhones, or laptop computers. Back then, members of Congress stayed in Washington on weekends catching up on legislation, golfing, playing cards, and otherwise socializing with each other, sometimes at a Washington Senators baseball game. Their children went to the same schools and churches and played intermural sports together.

They were a friendly, bipartisan mix whose focus revolved around national interests of import, and their lives and dialogue reflected it. Back then, they sat as a Congress five days a week, and each member was given two free round-trip airline tickets to come to Washington and return home after adjournment. Today, they are given fifty-two round-trip tickets and meet only Tuesdays through Thursdays and spend four days in their home districts or states raising money for the next election.

The best example I can give of their bipartisan spirit was the Halleck Clinic, and right down the hall was Democratic Speaker Sam Rayburn's Board of Education. Both were similar in function. After daily adjournment, the Republicans and Democrats convened on invitation and would debate the current day's legislative issues and conflicts over drinks, cigarettes and cigars, and card games. Both of these two private chambers below the Capitol rotunda consisted of three rooms and a toilet room. Each of the front rooms had a big round table that sat eight to ten, with comfortable side chairs all around against the wall. The back room had a small oval table for six to eight, and the third room had a cot for Congressmen Halleck or Rayburn to rest.

The standing, unspoken rule was that if you were invited (whether you were a Republican or a Democrat) and had a part of some contentious issue, you did not leave until it had been resolved. Members in the two rooms often exchanged seats and wandered back and forth trying to reach reconciliation. Members of the Senate freely crossed under the rotunda, unseen by the public. That year, 1956, visiting senators included Barry Goldwater, John Mansfield, Richard Nixon, John F. Kennedy, and Lyndon Johnson, to name just a few.

I had a front-row seat just inside the door and quietly sat to take notes, run errands, fetch other members, deliver messages, and replenish the liquor cabinet from the local store where both members had accounts. Sometimes I would be posted outside the door to control uninvited entrances.

My eyewitness take-home lesson was how the members maintained fraternal, collegial order. Regardless of the national weight and merit of a bill, they never shouted or became enraged. Yes, they'd raise their voices to make a point, but they rarely swore and only occasionally pounded the table. They were gentlemen with each other regardless of party. Their focus was on the issue being debated, not on each other, and the security of the nation. It was the Cold War era. Nothing was personalized, nor did partisanship ever surface. They understood the rule when they accepted an invitation, and that's how issues were resolved peacefully. When both Charles Halleck and Speaker Sam Rayburn retired, that bipartisanship ended. Once cameras were installed in the gallery and jet air travel became readily available, fund-raising became of paramount priority four days a week back in their home states.

Today, the two parties keep separate counsel within their ranks, and senators no longer cross the Capitol rotunda to reconcile differences regardless of the party affiliation, except in Conference Committee. It's all about brinkmanship, polarization, and partisanship; speaking from the floor to the cameras in the gallery; and inserting items into the *Congressional Record* to later articulate to their constituencies the illusion of their importance. I was an eyewitness to history back in the mid-1950s and realized how good legislation was crafted in a bipartisan way. If only Congress would learn from history and do it again!

When the initial ESA of 1966 was under consideration by both the Congress and the Department of the Interior, it was ten years after my first introduction to Washington. In between, I returned to Indiana to attend Valparaiso University in Congressman Halleck's district, working as a night watchman for the county highway truck and supply storage yard—a job he arranged for me since my family had no money. It was just below Lake Michigan and suffered from late lake-effect snow. So my job was to call out highway road crews when the snow required clearing roads.

The understanding I had with Halleck was simple: graduate from Valparaiso and go to Indiana University to study law and then return to Indiana and get into politics. During summers and seasonal spring and fall breaks, I'd work for Congressman Halleck as an aide, chauffeur, messenger, and so on and stay close to Congress and Capitol Hill. After law school, I returned to Washington in 1964 to practice law under Colonel William A. Roberts, the head of a Washington law firm who was friends with Stewart L. Udall, the secretary of the interior from 1961 to 1969. In all their meetings, I carried Colonel Roberts's briefcase and took notes, with follow-up responsibilities. Secretary Udall was a leader in the environmental awakening of the 1960s known as the "Green Revolution" (which I'll elaborate on in chapter 2), and Colonel Roberts was a participant in this initiative. I was a sportsman, as were many in Congress, and I got to know them as friends.

The champion of the ESA was Congressman John D. Dingell Jr. from Dearborn, Michigan, with whom I became friendly from hunting ducks on the Eastern Shore of Maryland two hours from the U.S. Capitol. Dingell was the principal advocate of the 1966 law and its later refinement in 1969 (when it was amended to include imperiled foreign species as well as domestic ones).

I discussed the ESA with him frequently as he crafted the language for the 1966 and 1969 laws and again in 1972–1973, when it was totally rewritten. Dr. Lee Talbot was part of that dialogue as senior scientist from the Council on Environmental Quality. He was the White House point man working with Congressman Dingell. I knew Lee, as we were both members of the Cosmos Club, the Explorers Club, and the Boone and Crockett Club. At Interior, where I was comfortable in working with the staff, I got to know Buff Bohlen, Nathaniel Reed, and Douglas Wheeler, as well as Dingell's congressional counsel Frank Potter, all of whom were working on drafting the ESA and were therefore in regular dialogue with Capitol Hill and the White House. It's only as I've gotten older and gained some perspective that I realized I was witness to the birth of the ESA. But that realization didn't sink in until later when the battles began.

CHAPTER 1

The Early Controversies

IN THE EARLY PREDAWN TWILIGHT HOURS ON SATURDAY, SEPTEMBER 26, 1981, John and Lucille Hogg were awakened by the furious barking of the family dog Shep, whose doghouse was just outside their bedroom window. John assumed that Shep had a run-in with a porcupine that occasionally frequented their Lazy BV Cattle Ranch west of Meeteetse, Wyoming, and so he turned over and went back to sleep.

What John Hogg discovered later that morning was that Shep had a run-in with a strange-looking mink or weasel-like creature that he'd killed. The creature had gotten into Shep's dog food dish, and the dog bit it in the head and broke its back. John threw the carcass over the fence, but, on reflection, Lucille later retrieved it for the local taxidermist to mount because of its devilish appearance, with a raccoon-like black mask over its penetrating black eyes and nose. It had black legs and feet with long claws, a long black-tipped tail, and smooth, silky tan hair covering its body.

The Hoggs took the carcass into the local taxidermist in Meeteetse, Larry LaFranchi, where Lucille operated Lucille's Café, a restaurant and coffee shop. The taxidermist immediately recognized the carcass as that of a black-footed ferret, one of twenty-three species on the endangered species list long considered extinct. He immediately called the offices of the Wyoming Game and Fish Department in Cody, Wyoming, and the U.S. Fish and Wildlife Service (FWS) in Billings, Montana, which later confirmed that the carcass was an allegedly extinct black-footed ferret. Dr. Susan Clark was also notified the same day; she was an independent biologist searching for black-footed ferrets in the area.

This black-footed ferret, named "Lucille" in honor of Lucille Hogg, was found by Lucille and John Hogg's ranch dog, Shep, on September 26, 1981, on their ranch near Meeteetse, Wyoming. The discovery of Lucille revealed that the black-footed ferret was not extinct, causing a media sensation and ultimately leading to the species being saved. In subsequent years, the back-footed ferret came to symbolize many of the greatest tensions and triumphs of the Endangered Species Act. *Ryan Hagerty/U.S. Fish and Wildlife Service*

Word spread immediately in that small rural community, and the local Park County game warden, Jim Lawrence, came to retrieve the carcass before the federal officials arrived. He was five feet, eight inches with a slender build, dressed in Levi's and a red shirt and his official warden's jacket with his badge prominently displayed. He had a local reputation for his authoritarian attitude and for being aggressive with a quick temper. He was one of those people who was never wrong, even when proven so. He was a loner, had no friends, and was shunned by his fellow officers. Lawrence had a history of tangling with the local taxidermist.

WANTED

DO
NOT
KILL
OR
TRAP

DO
NOT
KILL
OR
TRAP

$250 REWARD
for Photograph & Information

REWARD CONDITIONS: The ferret is an endangered species and is protected by very stringent federal and state laws. The reward WILL NOT be paid for any ferret caught intentionally in traps or killed by the finder. The reward will be given to the person providing information leading to the discovery and verification of the existence of black-footed ferrets (*Mustela nigripes*) in Wyoming. Skins and skeletons of ferrets struck accidentally by cars and found along roads, reports of ferrets seen or photographs taken in an area where a representative of the "Ferret Search" project subsequently observes a ferret will qualify for the reward. A few ferrets have been seen in most parts of Wyoming in recent years. Ferrets eat prairie dogs and are usually found on or near prairie dog towns.

IDENTIFYING CHARACTERISTICS: The ferret is the size of a medium sized mink, about 18 inches long and 2.5 pounds. *Unique features* are *black face mask* and *black feet.* Do not confuse it with long-tailed weasels (no mask or black feet) or domesticated ferrets in pet stores.

CONTACT: Ferret Search, Box 2705, Jackson, Wyoming 83001; Telephone: (307) 733-6856 as soon as possible after the sighting.

"Ferret Search" supported by National Geographic Society and The National Academy of Sciences and others.

During the 1970s and early 1980s, when the black-footed ferret was feared to be extinct, Dr. Clark and other volunteers searched across the West for evidence of ferrets, distributing thousands of posters like this one as well as postcards. *Image of poster provided by Franz J. Camenzind, PhD*

LaFranchi refused to surrender the carcass, saying that it was a federal matter and that the feds were on the way to retrieve it. Herein lies the rub. When Congress created the Endangered Species Act (ESA) a decade before, it intended to preserve states' sovereignty over managing listed species, such as the black-footed ferret. But in reality, executive branch regulations and judicial branch rulings would reduce the states' role to mere recipients of federal funding for the ESA program. During the legislative enactment process in 1971–1973, states' wariness of the feds preempting their rights was a recurring discussion. Congress responded by including language in the ESA calling for consultation and "cooperation to the maximum extent practical" between state and federal governments.

The cooperation clause proved ineffective that day in 1981. Lawrence became angry, and a fistfight ensued. Lawrence was badly beaten by LaFranchi, who knocked him out and threw him out into the street. Bleeding profusely, Lawrence crawled to the local bank across the street and called for backup.

Soon, Ron Creason of the FWS out of Billings arrived and collected the carcass. It was flown to Denver and then to Washington, D.C., and the Smithsonian for confirmation and identification. Then Park County Deputy Sheriff Bob Spears showed up and confirmed all the details. LaFranchi was arrested and charged with misdemeanor assault and battery (not only had Lawrence suffered a concussion, but he also required sixteen stitches over his eye) and interfering with a police officer. He pleaded guilty and was fined $100 plus ten days in jail; he further paid for the cost of replacing Lawrence's jacket, which was so bloodstained that it was ruined.

In defense of the feds (and LaFranchi), Lawrence was a troubled man, haunted by demons. He had been drinking and had assaulted his wife. He later killed his son, whom he disliked, when he returned from a basketball ball game late one evening and then turned the shotgun on himself and committed suicide. So ended the initial rediscovery of the black-footed ferret in Meeteetse, Wyoming. The carcass was stuffed and is now on exhibit at the Fish and Wildlife Service Museum at the National Conservation Training Center in Shepherdstown, West Virginia. It is nicknamed "Lucille" after Mrs. Hogg.

Fists would continue to fly in Wyoming over ferret management as state and federal authorities tried to negotiate and establish protocols. The Wyoming Game and Fish Department captured nineteen Meeteetse ferrets for a breeding program at its Wildlife Research Center at Sybille Canyon. But the facility lacked the resources to keep ferret-threatening diseases at bay and to train ferrets outdoors before releasing them to the wild. The FWS grappled with how to oversee Wyoming's management of the ferrets given the ESA's demands.

It was new territory for both parties to negotiate and establish protocols. Their recovery planning sessions grew heated. Wyoming, like most western states, resented federal involvement at any level, coveting its historic sovereignty and management over wildlife. The federal government challenged Sybille Canyon's effectiveness, and meetings intended to focus on recovery efforts deteriorated into shouting matches, table pounding, and—according to eyewitnesses—even fistfights in the parking lot! It has taken years for the wildlife agencies in the American West to fully accept and cooperatively integrate with federal oversight and direction, and some still struggle to do so. Invading state sovereignty is still a very touchy subject today throughout the West.

In the case of the ferret, Wyoming gave up after eighteen years of struggle. In 2005, the captive breeding program relocated to the FWS's National Black-Footed Ferret Conservation Center near Carr, Colorado. Captive breeding has been a resounding success, producing more than 10,000 ferrets to date, with no ill effects from inbreeding. The first reintroduction of black-footed ferrets to the wild occurred in 1991 in south-central Wyoming in Shirley Basin, where 228 ferrets were released over a four-year period as an experimental population. Ferrets have been reintroduced in thirty-two additional sites, thirty-one outside of Wyoming and one other in Wyoming—at Meeteetse, where the species was first rediscovered in 1981. The Meeteetse population was established in 2016 under a statewide regulation that took five long years to negotiate, as Wyoming and federal officials continue to struggle with effectively implementing the cooperative promise of the ESA. This population has experienced problems with sylvatic plague, and periodic releases are ongoing to sustain it. Complete recovery and delisting of the black-footed ferret may be possible within the next ten years provided that funding necessary to implement plague controls can be secured. This would be a remarkable achievement for a species that was once thought to be totally extinct.

Captive-bred black-footed ferrets have been reintroduced to more than thirty locations, dating back to 1991. In 2016, they were reintroduced with great fanfare to the Meeteetse, Wyoming, area, where they were first found in 1981. Several additional releases have occurred in Meeteetse since then. This September 2017 photo shows the author participating in one such release. *Kimberly Fraser/U.S. Fish and Wildlife Service*

The black-footed ferret has become an iconic symbol of the Endangered Species Act due to its cute appearance, status as a uniquely American species, and highly public presumed extinction, rediscovery, captive breeding, and reintroduction. *Kimberly Fraser/U.S. Fish and Wildlife Service*

THE OWL AND THE PRESIDENT

The states' wildlife management systems and organizational structures were built on the premise that wildlife was theirs to manage at their sole discretion, with states actually believing they owned the wildlife. But the ESA federalized wildlife management based on six federal international treaties to provide its constitutional footing. Congress deemed the prevention of the extinction of species vital to the survival of humankind and of the ecosystems on which humankind depends.

Therefore, the ESA is a comprehensive scheme to preserve the nation's genetic heritage and its incalculable value. The bald eagle—the national symbol—faced extinction until it was listed as "endangered" and protected by the ESA. Today, the bald eagle has recovered and has been "delisted." Also saved by time spent in Earth's emergency room: certain breeds of falcons, parakeets, sparrows, deer, kangaroo, whales, sea lions, daisies, and sunflowers. One peer-reviewed study estimated that the ESA has directly prevented the extinction of "roughly" 291 species[1]—a feat that few Americans recognize. Equally unrecognized is the fact that all 1,668 species on the threatened and endangered species lists have been sustained and conserved by the ESA. Without the ESA, many would have gone extinct.

Despite these successes, the ESA triggers volatile reactions over more than just state sovereignty issues. Controversially, the ESA places the protection of wildlife, ecosystems, and biodiversity ahead of economic, commercial, and industrial development. Because habitat loss is a primary cause of species decline and habitat restoration is our best tool to arrest declines, wildlife management actions impact land use. Thus, the line between regulating species and regulating land can be perilously thin. This means that the ESA can be a powerful tool for those who are interested primarily in how land is being managed provided that a species that qualifies for ESA protection can be found close at hand. The northern spotted owl was just such a species, and it rose to controversial prominence shortly after "Lucille's" discovery.

Its habitat, the old-growth forests of the Pacific Northwest, is home to Douglas firs, Sitka spruce, and western hemlocks—some 1,000 years old and soaring to heights as great as 200 to 300 feet. With trunks commonly about ten feet wide and sometimes as wide as twenty feet, these mammoth trees were the lifeblood of the commercial timber industry in western Washington, Oregon, and northern California. Local mills— critical employers in countless small towns—processed them into cut lumber for construction across the nation and for export abroad. Forestry and related industries play a massive role in America. Today, they provide 1 million direct jobs and more than $55.4 billion in direct payroll; when indirect and induced economic activity is included, they provide a total economic impact of 2.9 million jobs and $128.1 billion in payroll.[2] The overwhelming majority of these jobs (nearly 2.5 million) are generated by privately owned forestlands, which make up 59 percent of the total,[3] but federally managed public lands play a key role as well, especially in the Pacific Northwest.

An example of a clear-cut within old-growth timber in the Pacific Northwest. *John Zada/Alamy Stock Photo*

By the end of the 1980s, environmentalists could make a strong argument that the nation was long overdue for a deep, penetrating national dialogue about natural resource values, government agency priorities, and environmental harms weighed against corporate profit motives. Instead, the combination of these issues erupted into a bitter, acrimonious debate about owls versus human jobs and timber-dependent families and communities.[4] Environmental groups working to change timber industry practices and shrink its footprint in the Pacific Northwest ranged from the mainstream (such as the Portland and Seattle Audubon societies) to the extreme (such as Earth First!). This was the era of activists camping out in trees and chaining themselves to heavy equipment in order to stop timber sales. Nothing was off-limits.

The northern spotted owl was petitioned for listing under the ESA in 1986 and 1987.[5] But in late 1987, the FWS determined that listing the species was not warranted.[6] Twenty-three environmental organizations filed a lawsuit challenging this decision, and the court ruled in their favor in 1988.[7] The owl was subsequently listed as a threated species in 1990.[8]

Protestors arrested in Willamette National Forest on March 31, 1989. *Gerry Lewin/ USA Today Network*

The decision to list the species attracted 23,255 comments in fewer than seven months, and more than 80 percent of them opposed the listing—a level of public attention and criticism that was then unheard of for the ESA.[9] Loggers and their families, neighbors, and supporters held huge rallies, often sponsored by their employers, featuring slogans such as "Save a Logger—Eat an Owl," "I Love Spotted Owls—Fried," and "Save the Trees—Wipe Your Ass with a Spotted Owl" as well as boxes of "Spotted Owl Helper," a play on the iconic Betty Crocker product Hamburger Helper.

Protection of the northern spotted owl (*Strix occidentalis caurina*) under the Endangered Species Act (ESA) marked a turning point in the growing Timber Wars in the Pacific Northwest in the late 1980s and early 1990s. Prior to the listing of the owl in 1990, environmentalists had been using litigation under the National Environmental Policy Act of 1969 and the National Forest Management Act of 1976 in attempts to slow and even stop timber sales and logging on public lands in the region. The listing of the owl gave them a new, more powerful tool in the form of the ESA, and numerous timber sales were soon enjoined by courts. The controversy exploded in the public consciousness of America and played a role in the 1992 election of Bill Clinton. Ever since, the owl has been a negative symbol of the ESA, with numerous other species seen as "the next spotted owl," including the Louisiana black bear and the greater sage-grouse. *John and Karen Hollingsworth/ U.S. Fish and Wildlife Service*

The truth was that timber production on public land declined for a variety of reasons, including an increasing supply from Canada and a decline in demand, first as the export market shifted from finished to raw timber and later as the economic challenges of the early 1990s caused slumps in major export markets such as Japan. Even before that and before the spotted owl listing, timber companies embraced automation as a way to increase efficiency and profits by reducing their workforces. As early as 1978–1982, for example, timber employment in Oregon and Washington declined by 30 percent, from 136,000 to 95,000, due to automation.[10] But the media, looking for simplistic, recognizable thirty- to sixty-second sound bites that would resonate with the public, loved the spotted owl narrative. They, in turn, reinforced the messaging promoted by the timber companies, their supporters in Congress, and officials of the George H. W. Bush administration: that the problems facing the industry were the owl, the ESA, and environmentalists, who allegedly had been selfishly waging a decade-long war against logging, loggers, and their families and timber-dependent communities.[11]

The level of controversy around the owl was so great that to this day, the species is used to describe negative impacts stemming from the ESA. Thus, in 1990, when the Louisiana black bear was proposed for listing less than a week after the owl had been listed, it was referred to as "the next spotted owl" and "the spotted owl of the south." Ten years later, the greater sage-grouse would be seen as "the spotted owl of the plains," illustrating the lasting power of the owl as a symbol of the ESA as a force *opposed* to human well-being. It is no surprise, then, that listing the owl only deepened the conflict over forest management. Throughout the Timber Wars saga, both the Bush administration and Congress appeared determined to do everything within their power to ignore the owl and keep on "getting out the cut."

President Bush simply didn't have the environmental awareness I had hoped he would when, in the first week of his presidency, I agreed to advise him on conservation matters. As a leading member of the Boone and Crockett Club, I, along with several other club leaders, was introduced to the president by our club president Tim Hixon, who was a close friend of President Bush and would even vacation with him occasionally. During the campaign, Bush talked little about conservation and wildlife specifically. We urged him to develop a plan for wildlife conservation since the Council on Environmental Quality (CEQ) hadn't done so. Bush admitted he'd overlooked such an initiative and asked whether our group would take the lead and develop such a plan.

They all looked at me since I was an attorney in Washington, D.C., and well versed in contemporary conservation issues. President Bush asked whether I'd take the lead and work with CEQ to develop such an initiative. I agreed and immediately formed a working task force of senior Washington conservation professionals. We worked diligently for six weeks and produced a wildlife agenda for the president.

After review, Bush asked us to reconvene with him and the head of CEQ, Michael Deland. We initially met in the Roosevelt conference room, and as we discussed the agenda we'd drawn up, the president repeatedly told us why he couldn't do certain things, totally discounting the potential of our well-thought-out initiative. It sounded like an echo of President Reagan's administration instead of the "kinder, gentler" approach Bush had campaigned on. His primary concern was the Army Corps of Engineers. I became upset and said to Bush, "You are the president, those generals that run the Corps report to you. You can tell them what to do." He smiled quietly and explained there were laws and regulations they had to follow and why he really didn't have the control over the Department of Defense I thought he had.

I persisted to try to understand his position, visibly frustrated with his response. My agitated frustration had become visible when I felt the gentleman on my right bumping my knee under the table to slow me down. I persisted, and that gentleman to my right finally whispered in my ear that I couldn't talk to the president that way. Bush overheard him and said he recognized the work we'd carefully done to prepare the report and my frustration with his repeated responses. That gentleman to my right was Bill Spencer, president of Citicorp and past president of the Boone and Crockett Club.

After the meeting, I had repeat follow-up meetings with Michael Deland often at CEQ and occasionally with the president in the Oval Office, but our efforts barely made an impact on the administration.

The "God Squad"

The Interagency Spotted Owl Committee (ISC), an expert panel established by Congress in the first year of Bush's presidency, produced a conservation strategy that would protect the owl by setting aside 7.7 million acres of owl habitat in large, contiguous blocks in order to avoid fragmentation, including 3 million acres that otherwise would have been subject to logging.[12] The ISC's report was scientifically unassailable but largely ignored by both the Forest Service and the Bureau of Land Management (BLM) in the executive branch, which wished to continue logging at far higher levels.[13] In fact, BLM Director Cy Jamison's strategy was not to accept the ISC's recommendations for BLM lands but simply to cut the proposed timber protections in half.[14]

Congress continued to seek a solution by convening expert panels. In 1991, the House Committee on Agriculture and the House Committee on Merchant Marine and Fisheries jointly appointed a group of experts known as the "Gang of Four." Their report offered fourteen different management options,[15] which together demonstrated that it was impossible to maintain high timber harvest levels while meeting legal obligations to protect the owl.[16] This was not something the Bush administration wanted to hear, and the Forest Service and the BLM each paid lip service to both the ISC and the Gang of Four reports while largely ignoring their recommendations.[17]

In June 1991, the FWS issued a biological opinion finding that forty-four planned BLM timber sales would jeopardize the continued existence of the northern spotted owl.[18] The BLM responded by filing a petition asking the congressionally designated Endangered Species Committee—known around Washington as the "God Squad"[19]—to exempt those forty-four sales from federal interagency consultation under the ESA. On May 14, 1992, the committee voted 5–2 to exempt thirteen of the forty-four sales.[20] Ironically, after Bush lost his bid for reelection, those thirteen exemptions were overturned by a court that found his White House had violated the Administrative Procedure Act in its communications with the committee and its staff: three committee members had reportedly been summoned to the White House and pressured to vote for an exemption.

The northern spotted owl became a factor in Bush's 1992 race against Bill Clinton. On the campaign trail, Clinton promised to convene a summit to tackle the northern spotted owl issue within ninety days of taking office.[21] Bush, in contrast, said that "it's time to put people ahead of owls" and that "all across the country we have a spotted owl problem."[22] Of the ESA, Bush flatly declared, "The law is broken, and it must be fixed."[23]

On winning the election, Clinton made good on his promise, convening the president's Northwest Forest Summit in Portland, Oregon, on April 2, 1993. The summit was a full-day meeting, attended by President Clinton, Vice President Al Gore, and members of the cabinet, with the morning session focusing on the needs of the timber industry and local communities; the afternoon session centered on the northern spotted owl, its biology, and its habitat. At the end of the day, President Clinton promised a plan to address the situation.[24]

The president's Northwest Forest Summit, held in Portland, Oregon, on Friday, April 2, 1993, featured an entire day of testimony on the northern spotted owl, the timber industry, and rural communities in the Pacific Northwest. Attended by President Clinton, Vice President Gore, and members of the cabinet, it marked the fulfillment of one of Clinton's campaign-trail promises and in turn led to the development of the Northwest Forest Plan. *Bureau of Land Management/Flickr*

Following a third scientific study, conducted by the Forest Ecosystem Management Assessment Team, the comprehensive Northwest Forest Plan was formally adopted on April 13, 1994. The plan covered 24.5 million acres of public land, setting aside 19 million acres, 77 percent of the total, to protect the northern spotted owl and the marbled murrelet (a small seabird that nests in old-growth forests that had been listed as threatened in 1992). Limited logging would be permitted on 1.5 million acres, and 4 million acres would be fully open to logging, allowing potential yields of just over 1 billion board feet per year, one-quarter the historic level within the plan's boundaries. The sharp reduction in logging on public land under the Northwest Forest Plan was intended to conserve the owl and murrelet while allowing higher levels of logging to continue on state and private lands. Additional components of the plan included stream buffers to reduce erosion and protect aquatic habitat and species, especially salmon, and $1.2 billion in economic relief for communities in the region, addressing everything from funding for essential services like schools and police to retraining and unemployment assistance for displaced timber workers.[25]

Like any good compromise, the Northwest Forest Plan left nobody happy. Logging continued, and legislative riders successfully passed by the new Republican congressional majority in 1995 mandated the logging of many thousands of acres that were meant to be protected.[26] Yet economic impacts of the listing were nonetheless felt sharply for many years. Production of timber from covered lands dropped by three-quarters. The true rate of job loss is contested, with no two studies in perfect agreement, but a consensus has emerged that around 30,000 to 32,000 timber industry jobs were lost, creating rural refugees.[27] Moreover, the official designation of "critical habitat"—defined in the ESA as specific areas essential to the conservation of the species—was ordered by the court in 1992 (just as the court had ordered the owl's listing in 1990) and has been revised, re-revised, litigated, and fought over ever since.[28]

The ultimate role of the northern spotted owl in the decades-long Timber Wars was as a proxy for old-growth forest and as a scapegoat for the decline of the timber industry in the region. It should be acknowledged that this occurred with willing acquiescence from all sides. For the media, protecting owls versus jobs was a convenient and compelling narrative. For environmentalists, the owl was both a charismatic avatar the public could relate to better than mere trees and a legal foothold that allowed them to leverage the National Forest Management Act, the Federal Land Policy and Management Act, and later the ESA into injunctions against logging. For the timber companies, the owl, the ESA, and environmentalists were convenient foils. They could be blamed for the jobs crisis and avoid timber companies acknowledging the role of their own capitalist greed, overuse of limited natural resources, falling exports, and increasing use of automation. And for politicians as well, it was politically expedient to blame the ESA rather than timber companies for economic suffering in their districts and states.

In a final irony, the northern spotted owl population has continued to decline due to competition from an invasive species, the larger barred owl, which has been moving into the spotted owl's range and outcompeting it for food and shelter. Today, the eventual extinction of the northern spotted owl is a very real possibility.[29] This situation raises a host of penetrating questions about the ESA: Can the owl be saved? If it can't be, should we try anyway? Or give up? How much should we manipulate an ecosystem for the benefit of an endangered species? For example, in the owl case, some have proposed lethal control of barred owls within northern spotted owl habitat. Could such intervention be justified? Morally? Fiscally? Do these questions have different answers for different species, situations, and threats? How can we draw lines? *Who* should draw those lines? These are the questions America must ask itself as it wrestles with the moral dilemma now posed by climate change, biodiversity loss, and extinction. The concept of "triage," of abandoning species that cannot be saved in order to direct more resources to those that can be, is foreign to the ESA and controversial in wildlife conservation circles. But the truth is that, like doctors in an emergency room, biologists and policymakers at state and federal agencies alike have no choice but to make decisions about which species will be prioritized and which will not be. Take, for another example of controversy, the snail darter.

"WHATEVER THE COST"

In 1967, as America's "Green Revolution" started, construction finally began on the Tellico Dam, about twenty-five miles southwest of Knoxville, Tennessee. It was one of sixty-eight dams built by the Tennessee Valley Authority (TVA), which was established during the Great Depression and functioned as both a public utility and a regional economic development agency. The Tellico Dam was conceived in 1936, but due to World War II, it was deferred in 1942—and finally approved in 1959. Had its construction begun sooner than the 1960s, perhaps it wouldn't have sparked the intense local opposition it immediately did.

Billed as a hydroelectric power project, in truth, Tellico was primarily a real estate development scheme. The TVA had plans for a model town, housing developments, and industrial parks around the reservoir. To this end, the agency condemned 42,999 acres of private land, far more than was needed for the planned 16,500-acre reservoir.[30] The river that would be turned into a reservoir was widely regarded as the finest cold-water trout stream in the entire eastern United States. Moreover, the valley at large contained two former capital cities of the Cherokee nation, ancient burial grounds sacred to the Cherokee people, 285 separate archaeological sites, and 300 small farms and homes, all to be swallowed up by the filling reservoir.[31]

In 1971, a lawsuit against the dam was filed under the new National Environmental Policy Act (NEPA), leading to a 1972 preliminary injunction. The injunction was lifted in 1973 after the TVA completed the NEPA process.[32] But that year, a new species of perch was discovered by University of Tennessee biologist Dr. David Etnier and his student Bob Stiles. They named it the "snail darter" because its diet was largely snails.

Word of their discovery was relayed to University of Tennessee law professor Zygmunt Plater, and he and his student Hiram Hill, along with Trout Unlimited member Joseph P. Congleton, petitioned for its listing as an endangered species, which was granted in 1975. Plater and Hill were so optimistic that they had found a lever powerful enough to stop the TVA that they filed suit under the ESA to stop construction of the dam.[33]

The snail darter (*Percina tanasi*), discovered in 1973 and listed as an endangered species in 1975, was the subject of the landmark 1978 U.S. Supreme Court case *Tennessee Valley Authority v. Hill*, which held that under the Endangered Species Act of 1973, the survival of a tiny species of fish was more important than a $116 million federal dam that had almost been completed. The species ultimately survived in other rivers and streams, and it was downlisted to threatened in 1984 and delisted in 2022. © *Joel Sartore/Photo Ark*

The district court found in 1976 that the Tellico Dam would indeed—in violation of the act—"adversely modify" and perhaps destroy the snail darter's critical habitat. But it allowed construction to continue because it was already under way.[34] In 1977, the U.S. Sixth Circuit Court of Appeals reversed the district court and ordered construction halted despite the fact that the $116 million dam was by this time structurally complete and the only ongoing work related to its reservoir.[35] This decision instantly got the attention of Congress, and the attention of the entire nation followed in 1978 when, on June 15, the U.S. Supreme Court ruled in *TVA v. Hill* that construction of the Tellico Dam must remain halted.

This aerial photograph shows the Tellico Dam and environs shortly after the dam was completed. At the bottom right, the completed Tellico Dam stands athwart the Little Tennessee River, which has not yet filled the future reservoir. At the top left stands the Fort Loudoun Dam, on the main stem of the Tennessee River. This hydroelectric generating station, completed in 1943, provided the ostensible justification for the Tellico Dam, which diverts water from the Little Tennessee River to the Tennessee River above Fort Loudoun, whereas the Little Tennessee's natural outlet is below the dam. This water transfer raised the hydraulic head at Fort Loudoun and slightly increased the dam's electrical generating capacity. At the top right, U.S. Highway 321 stands on raised pylons above the future outflow from the Tellico Reservoir to the main stem of the Tennessee River. Down the center of the image runs the earthen berm constructed to hold back the reservoir. Today, almost all of the land to the right of the berm is inundated, as are 16,000 acres in total. *Courtesy of the Tennessee Valley Authority*

In the words of Chief Justice Warren E. Burger, the "plain intent of Congress in enacting [the ESA] was to halt and reverse the trend toward species extinction, *whatever the cost.* This is reflected not only in the stated policies of the Act, but in literally every section of the statute" (emphasis added).[36]

The impact of *TVA v. Hill* on the ESA was immediate and profound. Opposition to the ruling found its strongest advocates in Representative John Duncan Sr. (R-TN) and Senator Howard Baker Jr. (R-TN), and congressional committee testimony relating to the snail darter case ran well over 1,000 pages. Ultimately, Congress recognized that the ESA's language, as interpreted by the Supreme Court in *TVA v. Hill*, was too rigid and unforgiving to be effective.

As written, the ESA set the stage for a potentially infinite number of irreconcilable conflicts between species and human activities, such as in the snail darter case. Congress's solution to this issue in 1978 was an amendment to the ESA that included the aforementioned God Squad provision—the cabinet-level committee of senior government officials known as the Endangered Species Committee (ESC), which would wind up proving ineffective and unworkable.[37]

Before the ESC ruled on the northern spotted owl in 1992, it had convened only one other time: after its creation in 1978 to consider not only the Tellico Dam but also the Grayrocks Dam, which was threatening migrating whooping cranes on the Laramie River in Wyoming. The God Squad met on January 23, 1979, and unanimously voted *against* an exemption for Tellico while granting an exemption for Grayrocks provided that certain mitigation actions were included in that project.[38]

Senator Baker and Representative Duncan were incensed and immediately began seeking a way for the Tellico project to proceed through a rider on other legislation. Three times in 1979, Baker and Duncan attempted to insert language into other legislation that would force the completion of the Tellico Dam. On the third try, they succeeded in having their amendment enacted, when in forty-two seconds on June 18, 1979, on the House floor, Duncan inserted the following language into a must-pass appropriations bill: "notwithstanding the provisions of 16 U.S.C, chapter 35 [the ESA] or any other law, the [TVA] is authorized and directed to complete construction, operate and maintain the Tellico Dam and Reservoir project."[39]

"With regret," President Jimmy Carter signed the bill because, according to his adviser Stuart Eizenstat, he feared the ESA "would be chopped to ribbons with amendments exempting a whole variety of projects and species. There was a real concern that we could win the Tellico battle and lose the Endangered Species Act war."[40] With that, the dam entered operation on November 29, 1979, subsequently extirpating the snail darter population in the Little Tennessee River.

For the ESA, things were never the same after *TVA v. Hill*. With the visibility brought by the Supreme Court, the way the Tellico controversy dominated the attention of Congress all through 1978 and 1979, and coverage in the press from coast to coast, America awoke to the reality of the ESA. This law, then just a few years old, was not just a feel-good attempt to tweak land-use and trade laws to protect iconic species. No, the law had *teeth*, and it protected *all* species, even a lowly perch that no one had ever seen before. And it achieved that protection by interfering with other land uses, including major public works projects that were years into development. If a $116 million federal dam was not safe from the reach of the ESA, nothing was. Many across America, in federal agency offices, in state capitals, and in boardrooms, found this situation deeply unsettling. And we are living with the fallout to this day.

For the snail darter, the story has a happy ending. The TVA relocated a small population from the Little Tennessee River into the nearby Hiwassee River. In the following years, more snail darters were transplanted to nearby streams, including the Nolichucky, Holston, and Elk rivers.[41] In addition, the species was discovered in other streams in the Tennessee River valley.[42] Far from being rendered extinct by the Tellico Dam, the snail darter endured in these other streams and thrived.

In 1984, it was downlisted from endangered to threatened, and in 2022, the species was declared recovered and delisted in response to a petition filed by none other than Zygmunt Plater, the crusading law professor whose efforts to save the species back in the 1970s had led directly to *TVA v. Hill*.[43]

The initial application of the ESA and the controversies it spawned gave many the impression the law was inflexible and incompatible with human needs and prosperity. Fortunately, today, this could not be further from the truth. After decades of conflict around the management of listed species, bold experiments in collaborative partnerships for wildlife conservation, and increasing legislative and regulatory flexibility under the ESA, the law works better, and we understand it better. The early controversies described in these pages, the formative traumas of the ESA, have been left behind by the progress of subsequent years, and yet its dark mythology lingers, primarily in the West. The story of how the ESA was born of a "command-and-control" mentality but traced an arc toward collaboration and inclusion points us toward the possibility of a future that will be brighter still.

CHAPTER 2

The Command-and-Control Generation and the Green Revolution

NO LOOPHOLES OR OUTS. *ALL* SPECIES WERE TO BE SAVED *REGARDLESS OF the cost.* When the ESA was drafted, few (if any) realized it would be so rigidly interpreted and universally applied—not only to the snail darter but also to obscure reptiles, insects, other aquatic and terrestrial species, and even plants.[1] Few imagined the upset that the ESA would cause human land use. Most people in Washington imagined it as protecting charismatic, iconic species, such as polar and grizzly bears, eastern gray wolves, buffalo, whooping cranes, condors, alligators, and eagles. The act's comprehensiveness and inflexibility proved surprising and controversial in its early years.

To understand the rigidity of the 1973 ESA and the absolute primacy it vested in the federal government, you need to understand the top-down mentality of the era and of the men who drafted the legislation. Generations younger than my own have a hard time grasping the command-and-control mindset of post–World War II Washington, D.C. Military sacrifice and discipline were ingrained in all of us who grew up during the war. I recall my family reuniting in Indiana after the war—and almost all my aunts and uncles in military uniforms. This was normal for the time. One of my uncles lost an eye in the war; another lost the use of one arm, while neighbors lost limbs or their lives—they never came home. You never forget things like that.

In my own small town, our high school principal headed up the National Guard's medical unit. When he needed to make his quota in 1956, he added a year to my age so he could enlist me. A year later, after starting college, I was transferred to the Army Reserve. I then spent eight years as an army reservist while I attended college and law school, steeped in a disciplinary mindset of following orders. Back then, Cold War spies were everywhere, and the order of the day was to be vigilant and to keep your head down and your mouth shut. You never discussed your job, and you just kept moving.

Lorene Baier, my aunt, so well exemplified that spirit in her secretarial work for Bendix Aviation in South Bend, Indiana, that the Bendix CEO recommended her to General Dwight D. Eisenhower when he called him for his support staff—which is how I came to know General Eisenhower personally and the mentality of this era so well. Aunt Lorene was on General Eisenhower's personal twenty-four-hour staff in the London bunker when he was supreme allied commander during World War II. Their shifts were twelve hours on and off. Following the war, General Eisenhower recommended Aunt Lorene to General MacArthur when he became supreme commander of allied powers that ruled Japan beginning in 1945 until Eisenhower's election as president in 1952. When General Eisenhower (he preferred "General" over "President") was elected president, he summoned several of his London staff—including Aunt Lorene—back to help him manage the country.

Aunt Lorene sat in the office to the left of the Oval Office, with her desk against the common wall she shared with the president. When her door was ajar or open into the Oval Office, she had direct eye contact with the president. I became page boy for Congressman Charles Halleck in the U.S. House of Representatives in 1956. If I was released early from my page duties in Congress, I'd ride the trolley down Pennsylvania Avenue to the White House. I'd go in the back door to my aunt's office and signal my arrival. Frequently, she was working late, so I'd eat in the navy mess downstairs and hang out in the library or map room on the lower level of the White House. But on occasion, we'd get dinner on the way home, or she'd cook, as my apartment was a half block from hers on Massachusetts Avenue, a short distance from the White House.

Congressman Charles Halleck and the other members of Congress treated the page boys like family—even like sons. General Eisenhower—just my aunt's boss in my eyes—treated me like a grandson, reminding me of my beloved grandfather back in Indiana. I had pretty free access to the Oval Office with my aunt's permission, or the general would see me and wave me in, always looking for the gossip and latest inner workings on Capitol Hill. He asked me to personally deliver handwritten or typed letters to Charles Halleck, and the reverse became common practice. I was a sixteen-year-old kid, acting as the private messenger between the president, the most powerful man in America, and his legislative man on Capitol Hill, who was getting laws through Congress to fulfill the president's agenda. These experiences made me quite comfortable being around important people like the president, the White House staff, and members of Congress and their staffs. I was on a first-name basis with many, and it has continued. I've discussed legislation and national issues with many over the years right up to the present day.

As I observed the lead-up to the ESA's enactment in 1973 as a Washington lawyer, I was keenly aware that its seven draftsmen in the administration and Congress were hardened World War II veterans, many of whom had seen combat. So were the successive presidents prior to its enactment: Eisenhower, Kennedy, Johnson, and Richard Nixon, who signed the ESA into law. Before becoming president, Nixon was a naval officer and then served as Eisenhower's vice president for eight years. Their entire mental focus was a product of the federal command-and-control mentality and discipline of that time. Into this zeitgeist exploded the Green Revolution of the 1960s.

Mollie Beattie, FWS director from 1993 until her untimely passing in 1996, once stated that "what a country chooses to save is what a country chooses to say about itself."[2] The 1960s and 1970s produced the largest number of conservation, environmental, and consumer protection laws ever realized in the United States during two centuries of the republic. An environmental awakening spread across America, and a silent majority of concerned citizens began to mobilize and speak up, as noted in Douglas Brinkley's book *Silent Spring Revolution: John F. Kennedy, Rachel*

Carson, Lyndon Johnson, Richard Nixon, and the Great Environmental Awakening.[3] The public made the quality of the environment a national issue long overlooked during World War II and the subsequent growth of suburban America.[4] This "Green Revolution" became the emotional and sociological chrysalis that produced the ESAs of 1966, 1969, and 1973.

The troops coming home from World War II and, a decade later, from the Korean War started families and sparked a housing boom across America—creating suburbs, new towns, and shopping centers. Because housing and jobs for the troops was the nation's top priority, bulldozers destroyed pristine wildlife habitat that no one regarded for its intrinsic value.

Pesticides and other chemicals developed during World War II to industrialize food production and control noxious weeds and insects were developed without appropriate research, and no one was held accountable for their harmful effects on the environment. These products polluted lakes and rivers, freshwater ecosystems, aquatic life, forests, and plants and crops and destroyed the marble facades of historic buildings.

Rachel Carson's 1962 book *Silent Spring* publicized the effects of pesticides and provoked a national dialogue.[5] The smog and dirty air in major cities—especially New York and Los Angeles—caused serious health problems. Noxious industrial pollution discharged from smokestacks was characterized by the National Wildlife Federation as "poison confetti."[6] Some pedestrians began to wear gas masks just to stay safe on public streets. Biologists began documenting vanishing wildlife species and those at serious risk of extinction, such as the whooping crane, California condor, trumpeter swan, Florida kite, Hawaiian and Aleutian goose, prairie chicken, Atlantic salmon and sturgeon, Great Lakes trout, blue whale, monk seal, American green turtle, Key deer, black-footed ferret, eastern timber wolf, and grizzly and polar bears.

A series of major environmental disasters in the 1960s and 1970s resulted from disregard of the environment before, during, and after World War II. Historian J. Brooks Flippen said that the 1969 Santa Barbara oil spill "shocked Americans, placing environmental protection on the front burner in a way it never had been before, turning a concerned public into an activist one."[7] That same year, Cleveland's Cuyahoga River, polluted by industrial wastes, caught fire for the thirteenth time, and on this occasion, *Time* magazine photographed it. Then came New York's "Love Canal" carcinogenic landfill revelation in the early 1970s and the Three Mile Island nuclear plant meltdown in Pennsylvania in 1979.

Pollution and acidic toxic air and acid rain were so bad in the 1950s–1970s in Los Angeles and New York City that it spawned a series of ghastly gallows humor jokes. *Art Worden/Herald Examiner Collection/Los Angeles Public Library*

Ever present in the nation's consciousness at this time was also the potential for nuclear war, with its attendant pollution and contamination. The magnitude of these disasters was grist for the new media technology of television, which brought all these horrors directly into the public's living rooms on a daily basis. The advent of television and heightened media coverage became the connective tissue further fueling the explosive period of the 1960s and 1970s.

Environmental organizations swelled in membership from 125,000 in 1960 to 1 million in 1970, 2 million in 1980, and 6 million in 1990.[8] To realize their legislative objectives, this new environmental community followed the playbook of the Vietnam War protestors and civil rights activists of the era. An astounding sixty-seven environmental and consumer protection laws were enacted between 1963 and 1980 (see appendix B) as a result.[9]

From my many conversations with Representative John Dingell back then, I knew before it was even implemented that the Endangered Species Preservation Act of 1966 wouldn't have the teeth to be effective to meet the public's growing expectations of government intervention. It had little real authority to ban the import of foreign endangered species or parts thereof. It couldn't stop the killing or interstate trade of endangered species. Moreover, its constraints prevented the quick action required to save certain species of whales, whose imperiled status was earning public sympathy. Dingell, a hardened World War II veteran, was not going to lose this war. He tried to address the shortcomings of the 1966 law with the 1969 Endangered Species Conservation Act, but a third campaign would be necessary.

NIXON'S "KNIGHTS-ERRANT ON A RESCUE MISSION"

While President Nixon personally had little interest in environmental affairs,[10] he and his staff recognized the substantial vote coalescing around concern for the environment as they prepared for his 1972 reelection race.[11] Twenty million Americans participated in the festivities related to the first Earth Day, led by Senator Gaylord Nelson, on April 22, 1970.[12] But the event's organizers refused to meet with the Nixon White House, calling it a "billow of smog."[13]

Nixon wanted to capture the support of this visible and vocal voter bloc. Both chambers of Congress were controlled by the Democratic Party, so he directed his domestic policy advisers John Ehrlichman and John Whitaker to pursue and support the creation of environmental protection laws regardless of which political party originated them.[14]

On February 4, 1971, Representative Dingell introduced his bill to amend the 1969 ESA. It was presented as a mere three-page amendment to the 1966 and 1969 acts, both seen as benign by the American public at large.[15] Protection would be accorded not only to species "threatened with extinction" but also to those "that are a rare species." Dingell's intent was to protect an endangered species and its habitat *before* its situation became critical and it was threatened with extinction. He was motivated in part by the plight of the whales and the still-deficient 1969 legislation, which failed to permit the secretary of the interior to act until species were officially listed as endangered—no protection was afforded to species that were merely "threatened." The first session of the 92nd Congress ended without any action taken on Dingell's bill.

However, the Nixon administration had taken an interest and wanted to work with Dingell and his fellow Democrats to draft an entirely new ESA to supersede the 1966 and 1969 acts. For this task, Nixon relied on his friend Russell Train, whom he'd appointed to run the new CEQ, created in 1970 for natural resource and conservation legislative and policy ideas. Train had a generous, outgoing demeanor, and he was very well spoken. Of average build and height, he was approachable and clearly enjoyed his assignment. He took it seriously as the initial chairman of the CEQ, an area he was well versed in. He'd been Nixon's campaign finance chairman and knew the Washington political scene very well and how to work it in a very collegial spirit. Earlier, he'd been a judge on the Federal Tax Court and a leader of the World Wildlife Fund (WWF) and was comfortable with world leaders and international affairs. He would go on to become the second administrator of the Environmental Protection Agency (EPA) from 1973 to 1977 and would return to the WWF as its president in 1978, serving in that role until he became its chairman in 1985.

The drafting group, sometimes referred to as the "Train Team," included Nathaniel P. Reed, assistant secretary for fish and wildlife and parks; Deputy Assistant Secretary E. U. Curtis "Buff" Bohlen; and Douglas Wheeler, also a deputy assistant secretary to Reed.

Dr. Lee Talbot was Train's chief scientist at CEQ and knew the extinction crisis very well. Talbot was intensely involved in consultation on the development of the ESA but more in the conceptualization of the bill.[16] He was a very composed, confident man, slender and athletic in stature. When he spoke, people paid attention because he was so knowledgeable and deliberate in speaking with a matter-of-fact delivery. Moreover, he was direct and precise in his commentary, with a serious demeanor, not wasting any words, whether at a congressional hearing or socially. Small talk was not Talbot's style.

Reed referred to Talbot as a "spark plug" in the collaborative process.[17] Working in parallel with Interior, Talbot had separately outlined and written a few conceptual provisions of an endangered species bill with a global view that became part of the basis for the team to begin their formulation of the administration bill.[18]

Talbot was an officer in the U.S. Marine Corps (1953–1954) during the Korean War. His 1963 PhD at the University of California, Berkeley, was on endangered species. Dr. Talbot was a distinguished scientist, ecologist, and geographer who came from a likewise distinguished lineage. Talbot's mother was an ethnologist, and both his father and his mother were biologists. Talbot's maternal grandfather, known as the father of mammalogy, was the renowned Dr. C. Hart Merriam (1855–1942), first director of the U.S. Biological Survey for twenty-five years (which is today the FWS). He was the son of a wealthy New York U.S. congressman and led a distinguished career that began at age sixteen when he was appointed as naturalist of the Hayden Geological Survey of 1872, which became the basis for the creation of Yellowstone National Park that year. He was a contemporary and close friend of Theodore Roosevelt and, later, John Muir, with whom he spent time afield.

Talbot had been with the International Union for Conservation of Nature (IUCN) early on in his career as a staff ecologist and then joined the Smithsonian Museum in 1966. From there, he was transferred to become senior scientist at the newly created CEQ (1970–1978). Talbot later joined the WWF as its conservation director and science adviser in Switzerland (1978–1980) before returning to the IUCN as its director general (1980–1983), where in 1964 he had authored the first *Red Book* listing the endangered species of the world. Talbot's research took him to 134 countries on 160 expeditions over five continents. He authored more than 300 publications, including scientific research papers, monographs, books, and popular articles.

Chairman Russell Train (left) and Chief Scientist Dr. Lee Talbot (right) at the newly created Council on Environmental Quality in the White House were President Nixon's representatives to spearhead the creation and introduction of the 1973 Endangered Species Act (ESA) in Congress. They worked closely with the Department of the Interior and Congress to champion the enactment of the ESA, signed into law by President Nixon on December 28, 1973. *Associated Press*

Nathaniel "Nat" Reed at Interior was a tall man, well over six feet, with a slender athletic build and an imposing presence, always impeccably dressed from his patrician background, with the grace of eastern establishment mannerisms. From his years in public service, his leadership presence exuded confidence, forceful and knowledgeable as he spoke in a commanding, resonant voice. He was one who did not invite argument or small talk. Mr. Reed was always all business, and colleagues approached him with deference.

Reed was very well respected for his credibility and very strong minded. He was an intense-looking man with a humorless demeanor. He inspired loyalty and demanded excellence of himself and others. He refused to be hemmed in by the protocol and formal policies of Washington, D.C., thinking, and he urged his staff to think independently. Much of that is seen in the final ESA language.

He spent much of his career as a celebrated activist protecting Florida's wetlands and at-risk species. Reed was Florida Governor Claude R. Kirk Jr.'s secretary of the environment and understood balancing the needs of commerce and industry while protecting the environment. The Reeds were part of George H. W. Bush's social circle, as both maintained homes in Greenwich, Connecticut, and the Jupiter Island Club, which the Reed family founded and developed. Reed's brother Joseph was President Bush's chief of protocol following his career with his mentor David Rockefeller at the Chase Manhattan Bank.

E. U. Curtis "Buff" Bohlen was also a product of the eastern establishment. He had a charming Boston accent, his demeanor was that of the consummate gentleman, and his graceful movements over his lanky six-foot height demonstrated his athletic fitness. He and his wife, also a devoted environmentalist and recognized poet, summered at their home in Maine. Bohlen was intense, a deep thinker who pragmatically grasped the breadth of the national problem and its complexities. His inviting demeanor was methodical and conciliatory, always probing for an understanding of the issues and solutions for the complexities of the species extinction dilemma. He had no pretense, was easily approachable, and listened very well in dialogue. His composure was a great foil to Nathaniel Reed's ridged formality, balancing out the drafting team

Nathaniel Reed was assistant secretary for fish, wildlife, and parks at the Department of the Interior (1971–1977) during the enactment of the 1973 Endangered
Species Act (ESA). A celebrated conservationist from Florida, Reed led the charge
at Interior to draft the endangered species bill together with his able colleagues,
Deputy Assistant Secretaries E. U. Curtis "Buff" Bohlen and Douglas Wheeler.
Reed, Bohlen, and Wheeler were the administration's representatives who testified
during the 1972 and 1973 congressional hearings in addition to Russell Train,
chairman of the White House Council on Environmental Quality. They collectively
helped shepherd the ESA bill through Congress. *Department of Commerce Collec-
tion/State Library and Archives of Florida*

nicely. Bohlen also thought outside the typical Washington conservative box and was supported by Reed for his enlightened thinking. A protégé of Russell Train and the African Wildlife Leadership Institute, he had an international focus, as did Secretary Reed. Hence, following his work at Interior and before that in the foreign service corps, President Bush would eventually appoint him assistant secretary of the environment at the State Department.

E. U. Curtis "Buff" Bohlen (pictured here), deputy assistant secretary for fish, wildlife, and parks at the Department of the Interior, and Deputy Assistant Secretary Douglas Wheeler were the primary authors of the Nixon administration's Endangered Species Act bills submitted to Congress in 1972 and 1973. Both testified during the 1972–1973 congressional hearings and championed the bill through Congress. *Copyright C-SPAN*

Bohlen's co–deputy attorney, Douglas Wheeler, likewise recognized the complexities of drafting a new ESA to overcome the weaknesses of the 1966 and 1969 laws. Wheeler made a great team with Bohlen as they together drafted the new bill. A product of Long Island and Duke Law School, Wheeler had a diverse background. He would go into North Carolina politics, organize the American Farmland Trust for the Rockefeller family and become its first president, work for the National Trust for Historic Preservation and the WWF, and become executive director of the Sierra Club as well as California Governor Pete Wilson's secretary for natural resources. Wheeler was the consummate attorney, skilled at conceptualizing complex issues and finding solutions to thorny problem areas. He was six feet two, of normal stature, and looked like the consummate Washington, D.C., attorney, professional in his presentation and very articulate in his thought process as he carefully evaluated a situation within the greater context of the national extinction dilemma. Wheeler, like Bohlen, saw the big picture and dissected its parts to address its challenges. Bohlen had initially drafted a twenty-page outline of the deficiencies in the 1969 ESA, detailing amendments to correct the weaknesses. He then conceptualized the ideas and rewrote it with Wheeler as a draft for a new ESA bill totally superseding the 1966 and 1969 laws.[19]

Both were very bipartisan. They never created roadblocks for the Democratic congressional majority, and they worked with them in a conciliatory manner. They formed a collaborative team effort, working from conceptual ideas to the final language. During the drafting process, Reed briefed pertinent members of the House and Senate leadership on the proposed legislation to begin building alliances. The Nixon White House wanted to fast-track environmental legislation, giving the public little time to react. The naturalist Dr. Alston Chase stoically characterized this effort as "knights-errant on a rescue mission . . . they would build an ark to rescue the world."[20]

What was striking about the team that wrote the ESA was their mutual shared orientation. All were consummate gentlemen from traditional, wealthy East Coast families with good educations from the best prep schools and universities. They were extremely smart individually and collectively. Hard work marked their careers, as did a bipartisan disposition. Secretary of the Interior Morton was so concerned about their collective eastern orientations that he transferred all of them during the summer to work out of Interior's Denver office, hoping to broaden their geographic perspectives and overlooking their immense international perspectives.

"THE SECRETARY *SHALL*"

Representative Dingell, the CEQ, and the Department of the Interior leadership that drafted the bill for the president all took a very top-down approach in defining and drafting the unilateral authority vested in the federal government by the legislation,[21] as all were a product of World War II's command-and-control mentality. All were veterans harboring a militaristic approach to how the country should be governed.

Throughout 1972–1973, the ever-diligent Dr. Lee Talbot served as a watchdog for the administration.[22] He, Dingell, and Dingell's counsel Frank Potter all shared an equal passion to create a law that was absolute and unambiguous in its language. Talbot and Potter separately maintained close vigilance to remove any safety valves or weasel words and last-minute edits and amendments that would give future regulators an opportunity to avoid a threatened or endangered listing (which the 1966 and 1969 ESA laws were replete with), thus making the legislation an absolute directive.[23] These included changing wording such as "the Secretary may . . ." to "the Secretary shall . . . ," and they removed all phrases such as "insofar as practicable" and "insofar as possible."[24] One author later characterized them as "co-conspirators."[25] Deputy Assistant Secretary Bohlen was also involved throughout the legislative process, ensuring the integrity and intent of the legislation and the administration's perspective.[26] When one follows the trail of the actual legislative history throughout 1972–1973, their collective but separate oversight and scrutiny were essential in maintaining the integrity of the ESA as initially written, which John Dingell and others have verified.[27]

Moreover, in addition to being war veterans, all participants were committed environmentalists, with enviable records of achievement in wildlife conservation nationally and internationally. They were scientists (and attorneys) educated before or just after World War II. Their training in the wildlife sciences and management was based on an ideology and philosophy developed in the 1920s and 1930s by the very first wildlife biologists educated in those earlier decades of the Progressive Era.[28] Their commitment and motivation was clear notwithstanding Nixon's purely political approach to the subject.

Congress was driven by its collective dedication to a moral commitment to protect the American people from a pending catastrophe. Throughout the dialogue over 1971–1973, phrases like "moral imperative" and "ethical responsibility" were used in congressional debates. Examples included a "moral insensitivity to life," "ecological myopia," "best interest of mankind," "we are tinkering with our own futures," "to protect man from himself," "we cannot destroy what we cannot replace," "greed and our blind desire for progress has enabled us to interrupt the balance and rhythm of nature," and "extinction is quite literally a fate worse than death." This dialogue illustrates the mind and intent of the members of Congress at the time, one of complete bipartisanship and dedication to the public interest. They prioritized the public's future over progress and greed, in contrast to today's Congress.

Thus, the ESA was the result of a Democratic Congress and Republican administration—bipartisanship contrasting with today's division and brinksmanship.[29] The thoroughness with which the members of both the House and the Senate subcommittees and committees reviewed and examined the successive drafts, amendments, and technical changes to the endangered species bills over the two-year period is a vivid example of how Congress worked following World War II. Each member took their job seriously and applied themselves to lawmaking in the national interest as their top priority, not fund-raising and reelection. After World War II and the defeat of a totalitarian Nazi regime, preserving our democracy and historic system of self-governance and lawmaking was of paramount concern to our elected congressmen and senators. Their approach was practical and pragmatic, not partisan, in intent.

On July 24, 1973, the Senate enacted its version of the ESA 92–0, with 8 abstaining. Then, on September 18, 1973, the House approved its version by a vote of 391–12, with 31 not voting. On receipt of the Senate-passed bill by the House, they had struck out the entire Senate version of the bill except for the number and substituted the language of Dingell's H.R. 37. The bill was then submitted to a conference committee that adopted most of the provisions of the House-passed version of the bill, with amendments incorporating parts of the Senate version. Hence, most of its text was H.R. 37, the legislation that had been carefully shepherded through the House by Dingell, Potter, Reed, and Talbot.[30]

On December 19, 1973, the Senate unanimously approved the final legislation with little debate covering only a page of the *Congressional Record*.[31] The next day, the House also approved the bill with 355 yeas, 4 nays, and 73 not voting.[32] The House and Senate adjourned for the holidays on December 22, expecting that the president would sign the law that his knights-errant had worked so long and hard to create with the Democrat-led Congress.

Washington is a flurry of activity from Thanksgiving until Christmas, and 1973 was no exception. Nightly Christmas parties throughout the town abounded. Add to this the fact that the Nixon White House was in crisis and Congress adjourned on December 22 that year. Due to these factors, the ESA of 1973, "the single most important environmental act passed in the 20th century,"[33] in the words of famed biologist and scientist Dr. Edward O. Wilson, almost fell victim to a "pocket veto"—the term for legislation still unsigned by the president ten days after its passage if Congress has adjourned.

The president's support staff was dwindling throughout 1973. The Watergate break-in and cover-up investigation widened, as chronicled weekly in the *Washington Post* by investigative reporters Carl Bernstein and Bob Woodward. The Senate Select Committee on Presidential Campaign Activities held thirty-seven days of Watergate hearings in May and had subpoenaed Nixon's covert taped conversations. Chief of Staff Bob Haldeman and White House Domestic Affairs Adviser John Ehrlichman were forced to resign in April 1973 for their roles in Watergate, followed by Attorney General John Mitchell (all of whom were convicted and did jail time) and counsel John Dean, who was fired by Nixon for being a whistleblower. John Whitaker, formerly the environmental staffer responsible for the ESA as deputy assistant to the president under Ehrlichman, had moved to Interior on February 2, 1973, as undersecretary of the interior. Nixon had fired the Watergate special prosecutor Archibald Cox on October 20, and his attorney general, Elliot Richardson, and deputy, William Ruckelshaus, had resigned in protest on the same day, which became known as the "Saturday Night Massacre." The Arab oil embargo was causing a national energy crisis, enraging the public. Chief of Staff General Alexander Haig and Henry Kissinger (then secretary of state and national security adviser) remained to give the appearance of an orderly White House evidencing the penumbra of presidential power.

The evidence suggests that Nixon psychologically had quit the presidency weeks before the ESA passed Congress, when his disgraced vice president, Spiro Agnew, resigned amid a bribery and corruption scandal and Gerald Ford replaced him. The Watergate investigation begun in 1972 was broadening to include suspicion of the president's involvement in the cover-up, and his meetings with his defense attorneys over his potential involvement with Watergate and impeachment increased. Nixon had much to be distracted about, and he was glum. His behavior at times had become irrational and dysfunctional but was quietly masked by his staff. In December, he had indiscriminately scrawled on a White House notepad "Last Christmas here."

To the nation and his staff's bafflement, late on December 26, 1973, Nixon, First Lady Pat, and his daughter Tricia, plus twenty-five family members, staff, and Secret Service, secretly boarded United Airlines flight 55 from Dulles International Airport in Washington to San Francisco and then drove to his home in San Clemente, California, at a total cost of $4,841. The president's decision to fly commercially, abandoning the security of *Air Force One*, was unheard of for the modern presidency, described as an environmental gesture to save fuel. The public was confused by Nixon's erratic behavior.

When the president's secret and unscheduled departure was publicly announced the morning of December 27, Lee Talbot became concerned. Had the president signed the ESA sent to him by Congress on December 21? Dr. Talbot surmised that the president, amid the holiday festivities and related distractions and his secret departure on December 26, had not. Talbot raced to the White House and, working feverishly with staff, found the bill unsigned in a pile of congressional paperwork that the president's staff had failed to put on his desk for signature before his surprise departure.

The ESA of 1973, along with the other paperwork left behind, was couriered out to San Clemente, and the president signed it on Friday, December 28, 1973, without the expected fanfare. The White House staff issued a perfunctory signing statement just three days before the legislation would have expired under the ten-day constitutional deadline. Thus ended a relentless six-year campaign to protect America's threatened and endangered species. But out West, where Nixon would retire after resigning the presidency and where his interior secretary worried that the knights-errant lacked familiarity, rebellion was brewing.

CHAPTER 3

Beyond the 100th Meridian

FOR ALL HIS FAULTS AS A MAN AND AS A PRESIDENT, I LIKED RICHARD Nixon. I'd gotten to know him during his eight-year tenure as vice president under Eisenhower and found his demeanor friendly. Moreover, he was friends with Congressman Charles Halleck, for whom I served as page boy in 1956. In fact, I led the Young Republican group in Indiana during Nixon's presidential campaign in 1960.

Like me, Nixon was from a modest background. He was born on a citrus farm in Yorba Linda, California, and after he resigned the presidency, he spent the rest of his life in the Golden State. He famously blamed the East Coast media elites and intellectual class for his political problems—he remains a political punching bag to this day. But ironically, the moment Nixon left office, the environmental legacy he had established—however calculated—would be jeopardized by irascible westerners, resentful and distrustful of the Washington establishment.

In 1974, the year Nixon resigned, the U.S. District Court for the District of Columbia ruled in favor of a lawsuit brought against the Department of the Interior's BLM by the Natural Resources Defense Council (NRDC). The NRDC successfully argued that under Nixon's National Environmental Policy Act, the BLM was failing to properly assess the impact of the livestock grazing permits they granted ranchers to use public lands. Indeed, evidence showed that 84 percent of the land that the BLM oversaw was in only fair condition at best—much of it had deteriorated to "bad" or "poor" conditions. So Interior agreed to reduce the intensity of grazing on most of its lands—which in the West is 30 to 70 percent of all the land, depending on the state.

The Ford administration brought further insult to westerners. The 1976 passage of the Federal Land Policy and Management Act (FLPMA) ended the federal government's century-old policy of selling or giving away western public lands. The new policy is to retain, protect, and manage the remaining western lands not already disposed of or incorporated into national forests, national parks, and so on—mostly low-quality rangelands remain. The BLM is directed to manage these lands, with FLPMA serving as a new organic act for the agency.

By 1977, when newly elected President Jimmy Carter launched an ill-fated attempt to cancel thirty-two planned water projects, alienating western senators of both parties and western water users alike, a regional political realignment was well under way.[1] In the mid-1970s, Republicans began defeating longtime Democratic incumbents in Nevada, New Mexico, and Wyoming. The most shocking upset was the defeat of three-term incumbent senator Frank Moss of Utah, a strong supporter of federal wilderness protections. Orrin Hatch, an attorney who was born in Pittsburgh, Pennsylvania, and who had never held office, handily defeated him in 1976, and one of his first acts as a senator was sponsoring legislation to transfer federal lands to state ownership and control.

While Washington continued introducing and debating legislation to address the concerns of the Green Revolution, out West the so-called Sagebrush Rebellion blazed in opposition to Carter's preservationist policies. In 1979, Utah used a government road crew to physically breach a BLM-controlled wilderness area and build a road through it. (This incident began a long tradition in Utah of spending government resources to oppose federal land policies.) The same year, Nevada passed a law claiming title over BLM lands within its borders. Arizona, New Mexico, Utah, and Wyoming soon followed with their own legislative actions—and Wyoming went so far as to claim title over Forest Service lands as well.

Contrast these behaviors with how well environmental skirmishes in the East over ESA enforcement were settled during this same era. The Marine Mammal Protection Act of 1972, another Nixon-era law, caused economic hardship to those who made a living selling scrimshaw—traditional American folk carvings on the teeth and bones of the endangered sperm whale—by prohibiting the import of marine mammals' parts for commercial purposes. Dating back to the late 1700s, the art of scrimshaw was widely practiced by sailors aboard American whaling ships who had both ready access to whale teeth and leisure time in which to hone their craft. Most closely associated with the whaling communities of New England, scrimshaw became a popular example of American folk art. The decline and end of commercial whaling in the twentieth century also led to the decline of scrimshaw, and examples are now seen primarily as historical artifacts.

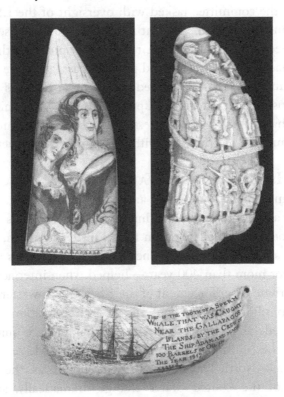

The art of scrimshaw, carving the teeth and bones of whales (and sometimes other ivory) into images and objects, dates back to the late 1700s and became popular after 1815. It was widely practiced by sailors aboard American whaling ships. The pieces shown here illustrate a variety of styles. *Courtesy of the New Bedford Whaling Museum*

Not only did scrimshaw artists and businesses throughout New England suffer as a result of the 1972 law, but the 1973 Endangered Species Act also wiped out the value of the large inventories of raw materials they had invested in because it banned interstate commerce in those items. (The Marine Mammal Protection Act had only restricted their international trade.) Inflation in 1972 and a recession in 1973 worsened their economic situation. But by working with Congress instead of rebelling, they were able ameliorate their situation through legislation.

The House Merchant Marine Committee was populated by members of Congress from eastern states with marine interests and at this point in time was the committee tasked with oversight of the ESA. In 1976, Congress completed an amendment to the ESA allowing scrimshanders to receive special permits to sell their scrimshaw provided that the whale parts used had been lawfully owned prior to passage of the ESA. In the same amendment, Congress allowed the General Services Administration to dispose of its strategic stockpile of more than 23 million pounds of sperm whale oil by sale. Congress thus resolved two major conflicts between commerce and the ESA without any adverse impacts on species.

"WHISKEY IS FOR DRINKING, WATER IS FOR FIGHTING OVER"

Wildlife management is not about animals and plants alone; it's 10 percent science and 90 percent managing people's interactions and expectations. The cooperative collaboration found in the southern, midwestern, and eastern states—learned skills necessitated by people's proximity to each other for more than 400 years—helped when interfacing with the federal government's enforcement of the ESA. But to westerners, accepting the responsibilities of the ESA and other preservation laws imposed by a distant federal government in Washington, D.C., is anathema.

West of the 100th meridian—the line of longitude running roughly from Bismarck, North Dakota, to Brownsville, Texas, the climate and topography of the United States change dramatically, and the culture reflects it. Except for the coastal stretches of California, Oregon, and Washington, the West is mostly wide-open spaces, twisted scenic beauty, and an unforgiving climate of parched summers, drought, and brutal winters. It is an arid land of alkali flats where little lives besides sagebrush—after which the "rebels" named themselves. Much of it was and is incapable of being tamed and cultivated and growing crops.

The West's rural population has always been spaced far apart, separated by mountains and wide-open rangeland instead of neighborly white picket fences. Early western settlers didn't have to learn about cooperation and collaboration with each other. Individualism and self-reliance were their learned skills after their one-way wagon train trip from the East and their lives as lonely, restless homesteaders began. Collaboration was never in their DNA. When forced by the ESA to work cooperatively with the federal government, westerners resisted its burdens; many simply ignored it and followed the rule of "shoot, shovel, and shut up."

I have personal knowledge of the history and characteristics of westerners and a visceral understanding of their resentments. For more than three decades, I owned a small ranch in Montana where I lived full-time during the recession that ultimately cost President George H. W. Bush reelection in 1992. My property was not far from my late grandfather's homestead, which had been gobbled up by a nearby ranch after his death in the late 1950s. From 1946 until 1950, my summer vacations from school consisted of my taking a train to meet him—as an elementary school boy, all alone from Chicago to Billings, transferring in Minneapolis under the watchful eye of a railroad conductor whom my father would tip.

My grandfather Peter Baier had a farm in Indiana that his wife and children would work in his absence. He was a German immigrant with a wanderlust to travel, hunt, and fish, so, in 1915, he staked a homestead fifty miles southwest of Billings, Montana. Each year, after planting his spring crops in Indiana, he'd drive out West in his old touring car and return just in time for the fall harvest.

Grandpa Pete and I would stay in the small log cabin he built just south of Luther, which was a crossroads with a small general store and post office combined. The cabin had no electricity, heat, or running water. It had an outhouse, and an adjacent spring provided fresh water. Like much of the land in the West, my grandfather's homestead was really unproductive land, half of it running up the west side of the Beartooth Mountains. When he was out West, he'd make his living by selling

Watkins products—salves, tonics, lotions, liniments, and assorted health aids for animals and humans. He was the next generation of a legitimate snake oil salesman, but it got him traveling around Montana, Wyoming, and the Dakotas, camping in an old army tent on the land of his customers, many of whom were second- and third-generation homesteaders. Some were Indians, whom my grandfather loved and with whom he would barter Watkins products for beaded artwork that I still possess and treasure. We visited late into the night listening to their stories. Sitting and listening to these stories, year after year, gave me a unique insight into what life in the West used to be like.

American Progress by John Gast (1872), depicting manifest destiny of the people moving westward, guided and protected by Columbia, representing American and classical republicanism. The settlers are aided by the technological innovations of the railroad and telegraph, driving Native Americans and bison into obscurity. Columbia is bringing the "light" (signifying enlightenment) from the eastern side of the painting toward the darkened west. *Library of Congress*

I learned it was not the "Old West" popularized by Hollywood, Madison Avenue, pulp fiction novels, John Wayne, and Clint Eastwood. The traditional Old West was one of toil, desperation, grief, and opportunism as the early homesteaders in their covered wagons and oxen struggled to travel to the West over inhospitable rugged terrain and impossible weather conditions. Turning the desert they found into a garden was a myth sold to them by Congress and the Homestead Act, the intention of which was to colonize the West and achieve American "manifest destiny" by transferring ownership of the land from the Indians to pioneers. Congress thought it worked well in settling the Midwest, so why not in the West?

The government's promise of freedom and prosperity was a paradox of scarcity. Much of the land claimed by homesteaders was arid, far from water and markets and incapable of sustaining life or encouraging entrepreneurial ownership.

But these pioneers had accepted a one-way fate. Wagon trains only went west, not east. And unlike my grandfather, who could drive his car back to his fertile midwestern soil every fall, the early homesteaders were forced to rampantly exhaust the natural resources around them and then move on to another location to sustain themselves. They seldom stayed in one place and established an identity, instead always looking for more productive land. To quote Ralph Ellison, "If you don't know where you are, you probably don't know who you are."[2]

The West's regional character of isolated, lonely, rugged individualism contrasts sharply with populations who have a sense of place from which a responsible democratic land ethic develops and is sustained by collaboration, cooperation, and community. The angst and restlessness of early homesteading has made a deep imprint on westerners, explaining their historical disregard for the environment and their distrust of the government.

Left on their own in such arid conditions, westerners treated natural resources—grass, timber, and especially water—as though an inexhaustible supply of these resources existed, despoiling the land that had once given them hope. That barren land denied their hope and shaped their lives and values. The settlers ascribed to the land they owned the

American ideals of freedom, opportunity, material success, and manifest destiny. But this psychological survival mechanism was no replacement for what they lacked: physical and spiritual bonds, lifelong friendships, institutions, and history, memories that develop in a community and stable society. They failed to submit and adapt to the limits of nature by allowing it to help shape their way of life. They dictated their own terms to nature. This is the essential key to understand why the West, which thinks of itself as a citadel of self-reliance, has presented such resistance to the enforcement of the ESA.

In the "new West" of today, you can hear voices of the past as clearly as you can see the deep ruts in the trails of the old covered wagon trains. Today's in-migration is mostly from California, Texas, and the East and is homogenized by the media's mythic West of urbanized cowboys. Today's migrants have driven up land values far beyond their worth for agricultural uses. Aridity, droughts, and wide-open spaces remain, and 75 percent of the population live in cities, overpopulating and outstripping their water resources.

Hydraulic engineering; large-scale drilling for water, oil, and gas; timber harvesting; and mining exploit the land and its natural resources, pressing the West's limits of nature just as the first homesteaders did. Fueled by capitalists, Wall Street, hydraulic technology companies, and government power elites, the prevailing attitude today is that natural resources are inexhaustible. Both new and old westerners still refuse to adapt to nature's limits, continuing to believe that high-tech hydraulic engineering and technology will ensure their continued domination and meet their ever-expanding population's demands.

Westerners' schemes have depleted rivers like the Colorado, Green, Rio Grande, and Platt, reservoirs like Powell and Mead, and even the Great Salt Lake. As the old saying, often attributed to Mark Twain, goes, "Whisky is for drinking, and water is for fighting over," and this idea has been borne out.

A SAGEBRUSH REBEL IN THE WHITE HOUSE

The best-known champion of this western mindset was presidential candidate Ronald Reagan, a former California governor who on the national campaign trail in 1980 declared himself a "Sagebrush Rebel"[3] as part of his small-government, pro-business persona. Reagan ran on a platform of deregulation, states' rights, resource production, prosperity, and American greatness.

When Reagan was swept into office by a landslide win over incumbent Jimmy Carter in 1980, it seemed that a major reaction against environmental protection, including wildlife and wildlife habitat protection, was close at hand. An ominous sign was his appointment of James Watt as secretary of the interior. Watt was recommended to Reagan by another Sagebrush Rebel, Nevada Senator Paul Laxalt.

Reagan himself graciously listened when a few of my Boone and Crockett members and I voiced our vehement concerns to him during several small meetings in the Oval Office, but he remained noncommittal, a true politician and actor to the end. I don't think anyone ever got to know Reagan as a friend, as detailed by his biographer and my friend Edmund Morris in his book *Dutch*.[4]

Thankfully, however, natural resources issues were of little interest to Reagan personally, who was more focused on national security, foreign policy, and taxes. Moreover, Watt and Anne Gorsuch—Reagan's appointee at the EPA—grossly overplayed their efforts to reduce environmental protections and were soon forced out, replaced by more traditional, moderate Republicans, such as William P. Clark and then Donald P. Hodel at Interior and William Ruckelshaus and then Lee Thomas at the EPA.

In the view of historians including Otis Graham, James Turner, and Andrew Isenberg, Reagan's environmental policy was ultimately a failure. The changes made early on by Watt and Gorsuch were fleeting. In exchange for fewer acres of parks, a few more coal leases, and a few years of lean environmental budgets, the administration greatly strengthened the hand of the largest and most active organizations in the environmental community by helping them attract donors. As chapter 5 will detail, these organizations would later flex their muscle to influence the administration of the ESA through lobbying and litigation, and they would be more influential in shaping policy than the Reagan administration ever was.[5]

James G. Watt (1938–2023) served as secretary of the interior for the first three years of Reagan's presidency (1981–1983). He promptly and indelibly established himself as the greatest antagonist to the Endangered Species Act (ESA) ever to serve in the federal government. Watt saw as his mission the exploitation of the nation's natural resources and made it clear that he viewed extractive industries as friends and environmentalists as enemies. When Watt took office, listings under the ESA immediately ground to a virtual halt, prompting congressional scrutiny and new statutory deadlines in the 1982 amendments to the ESA, designed to hold Interior's feet to the fire in its application of the ESA. These deadlines, ironically, later became a potent weapon in the hands of environmental extremists on the left, who were themselves able to grind the ESA listing program to a halt by overloading it with listing petitions in the late 2000s. Watt was unceremoniously fired by Reagan after his generally abrasive demeanor and particularly offensive remarks directed toward minorities and those with disabilities became too much to bear.*

---

* David Hoffman, "Watt Submits Resignation as Interior Secretary," *Washington Post*, October 10, 1983, https://www.washingtonpost.com/archive/politics/1983/10/10/watt-submits-resignation -as-interior-secretary/84ba758c-03f2-439d-8105-0bab802247b9/.

Ironically, the "Sagebrush Rebel" president signed into law two major amendments to the ESA during his two terms in office. A 1982 amendment passed both the House and the Senate on voice votes, and a 1988 amendment passed the House by a vote of 399–16 and the Senate by a vote of 93–2. Neither of these immensely consequential acts of Congress was accompanied by any of the hallmarks of legislation familiar in today's divided government. There was no head counting, no last-minute vote whipping, no filibuster threats, and no anxious waiting to see whether holdout legislators would come around. The partisanship that we know today had not yet gripped America.

The greatest legacy of these amendments were unquestionably two totally new provisions that vastly expanded the flexibility of the ESA for states, landowners, and regulated commercial entities. Before the amendment process, the impact of ESA regulations on state wildlife management and private landowners was such that state fish and wildlife agencies, zoos, and other partners struggled to find suitable locations to reintroduce captive-bred populations of species—such as the iconic whooping crane—because of conflicts with other land uses.

The amendment process established a provision known as "experimental populations" that facilitates reintroduction of listed species *outside* of their existing habitat. With experimental populations, Congress authorized reduced protections for introduced populations, further customized to meet the needs of specific species and locations through individual regulations. This development made the ESA more flexible and less conflict prone and facilitated the reintroduction of species such as the whooping crane, the California condor, the black-footed ferret, and the Mexican gray wolf into new areas.

Even more consequentially, Congress provided a means to authorize incidental take of listed species. Prior to the 1982 amendments, the ESA was absolute in its prohibition on take of any listed species, including either killing or injury. (The ESA defines "take" as "to harass, harm, pursue, hunt, shoot, wound, kill, trap, capture, or collect, or to attempt to engage in any such conduct.") This absolute prohibition led to the frustration and near failure of a five-year collaborative effort on San Bruno Mountain on the shores of San Francisco Bay just south of San Francisco.

The whooping crane (*Grus americana*) has been the subject of intensive conservation efforts, including captive breeding, since the 1940s, long before the species was listed as endangered in 1967. The species' population has increased from a low of sixteen in 1941 to 836 today. *Joe Ferrer/Shutterstock*

Early September is typically one of the warmest times of the year in San Francisco, and September 1982 was no exception. The towering forty-five-story glass-and-steel Four Embarcadero Center had just been completed that year, and it was in one of its many conference rooms that the future of the ESA transformed from one where conflict was inevitable to one where private landowners and the ESA could coexist. Around the table sat twenty-five to thirty frustrated and exhausted negotiators, including attorneys representing real estate developers; local activists from San Mateo County, California (just south of San Francisco); San Mateo County officials; representatives of the cities of Brisbane, Daly City, and South San Francisco; the California Department of Fish and Game; and the FWS, including both attorneys and regional leadership who had flown in from their offices in Portland, Oregon.[6]

The group consisted of proponents, negotiators, former adversaries, friendly rivals, and occasionally bitter opponents who had, through three years of effort and many difficult lawsuits, negotiations, and compromises, slowly and painfully pulled together an unprecedented conservation plan. Through years of difficult negotiations, they had agreed to preserve open space, conserve imperiled butterflies (specifically the Mission blue butterfly, the San Bruno elfin butterfly, and the callippe silverspot butterfly), fund long-term conservation, and assuage the worries of a long-anxious community, all while carrying out a massive, multi-million-dollar real estate development on the last significant unprotected open space near San Francisco. The location: San Bruno Mountain, a green and brown edifice rising above San Francisco Bay and cutting halfway across the San Francisco Peninsula. Surrounded by dense urban development except on its eastern side facing the bay, it was recognized as a global biodiversity hot spot by none other than legendary biologist Dr. Edward O. Wilson.[7]

View of San Bruno Mountain State and County Park, San Mateo County, California. Home to several species of imperiled butterflies, San Bruno Mountain provided the model for the development of habitat conservation plans in the early 1980s. *Courtesy of the County of San Mateo Parks Department*

Three species of imperiled butterflies live on San Bruno Mountain. From top to bottom, the endangered Mission blue butterfly (*Icaricia icarioides missionensis*), the endangered San Bruno elfin butterfly (*Callophrys mossii bayensis*), and the endangered callippe silverspot butterfly (*Speyeria callippe callippe*). *Photographs by (from top to bottom) U.S. Fish and Wildlife Service National Digital Library, Donna Pomeroy, and Patrick Kobernus*

The comprehensive plan created by the group between 1980 and 1982—the San Bruno Mountain Area Habitat Conservation Plan— would protect 90 percent of the mountain and 86 percent of its butterfly habitat. It addressed not just the butterflies but also fifty-four different species present on the mountain—some listed under the ESA, some not. The plan featured phased construction projects, perpetual protection for some habitat, an ongoing management plan, and a sustainable fee structure to fund habitat restoration and other management activities. Developers would pay start-up fees, and thereafter residents would pay annual fees.[8]

The problem the group faced was the inflexibility of the ESA because "the Endangered Species Act prohibits killing or injuring any endangered butterfly. No development [on San Bruno Mountain] could occur on private or public land without killing or injuring some of the butterflies, and therefore, at present, no development is possible,"[9] their plan noted. After considerable debate, lawyers in the group concluded that ESA permits for "enhancement of survival" were not available for commercial development projects like theirs, and they were at an impasse. Only Congress could break the logjam.

Fortunately, Congress was considering the 1982 amendments to the ESA, and those present were well connected on Capitol Hill. The entire group used the conference room speakerphone to place a call to Steve Shimberg, counsel to the Subcommittee on Environmental Pollution of the Senate Committee on Environment and Public Works, who was managing the ESA legislation, then before a joint House-Senate Conference Committee. Shimberg was in Washington, D.C., three hours ahead of San Francisco, but it was crunch time, and he was working late; Shimberg directed his staff to begin drafting implementing language that very evening.[10] The amendment that came out of the Conference Committee was a legislative endorsement and codification of the San Bruno Mountain plan. The conference report even cited the San Bruno Mountain Habitat Conservation Plan by name, holding it up as the model for future habitat conservation plans (HCPs) because it was based on extensive biological studies, preserved significant amounts of open space, and was adequately funded.[11] Under the final 1982 ESA amendment, incidental take permits (ITPs) may be issued for any taking "incidental to, and not the purpose of, the carrying out of an otherwise lawful activity." They must be supported by an HCP that identifies the impacts of the taking, demonstrates that the impacts will be minimized and mitigated as far as possible (and that there is funding for this), and explains the other alternatives that were considered and rejected (and why). For the first time ever, there was a legal mechanism to authorize incidental take of listed species, sparing landowners from the draconian take penalties imposed under the heretofore ironclad ESA. This compromise saved the San Bruno Mountain project and would later form the basis for an increasing number of innovative approaches to deconflicting development and the ESA. Flexibility, collaboration, and compromise would be the new keys to endangered species management.

THE PROPERTY RIGHTS WARS

The democratic process was working in the 1980s, and the ESA was changing appropriately through legislation, as its creators intended. For proponents of the Sagebrush Rebellion, the Reagan administration was a disappointment in Washington and an unfulfilled promise. Reagan notably did not campaign on the Sagebrush Rebellion when he ran for reelection in 1984; he had ample other issues to run on, including the strong economic recovery and increasing international prestige of the United States.

The original Sagebrush Rebellion was effectively over at this point, but it expanded and matured into a broader pro–property rights movement that spanned the nation, advanced by powerful lobbying interests, such as Citizens for a Sound Economy, and savvy legal operatives, such as the Pacific Legal Foundation and the Mountain States Legal Foundation (initially led by James Watt before Reagan appointed him secretary of the interior).

In 1988, an opaquely named "Multiple-Use Strategy Conference" was held not in Washington, D.C., but in Reno, Nevada. This three-day meeting brought together individuals, public interest groups, trade and government organizations, and industry (mining, fisheries, timber, petroleum, and agriculture) united in their view that economic development and human prosperity are compatible with a healthy ecosystem and that the environmental movement is unnecessarily hostile to economic prosperity. At this meeting, attendees launched the Wise Use Movement and adopted a twenty-five-point "Wise Use Agenda." The agenda called for the opening of more public lands to grazing, mining, and forestry; recognition of enhanced private property rights; legal impediments to environmental litigation; and the weakening of the ESA, among other things.

The northern spotted owl saga, as recounted in chapter 1, poured gasoline on the fire after Reagan's time in office ended. The property rights movement surged, with an explosion of short-lived but influential movements similar in orientation to the Wise Use Movement. They emphasized individualism, private property, and protection of individual constitutional rights at the expense of traditional environmental laws, such as the ESA, the Clean Air Act, and the Clean Water Act.[12] In 1990, Catron County, New Mexico, passed government ordinances to interfere with federal enforcement of environmental laws within the county. Eventually, forty-five counties in six states joined this populist "County Movement," although their actions lacked any legal justification and were largely symbolic. The federal government, however, was on notice that in the West, it had lost credibility and even authority.

Having refused to attend his trial for firearms charges, Randy Weaver was a fugitive from justice when in 1992 federal marshals approached his cabin in Ruby Ridge, Idaho. A firefight followed by an eleven-day siege ensued. Weaver's wife, fourteen-year-old son, and dog were all shot dead; so was Deputy Marshal Bill Degan.

In 1993, the Bureau of Alcohol, Tobacco, and Firearms, suspecting the presence of illegally converted fully automatic firearms, conducted a fifty-one-day siege of the Waco, Texas, compound of the Branch Davidians, an apocalyptic cult descended from Seventh-day Adventists. The siege led to the deaths of four federal agents—as well as eighty-two

Branch Davidians, including twenty-eight children. Timothy McVeigh later cited Waco as the reason for the Oklahoma City bombing in 1995, which he carried out on the second anniversary of the siege's end.

In 1994, Nye County, Nevada, Commissioner Dick Carver drove a bulldozer through a closed Forest Service road on July 4. A year later, his picture was featured on the cover of *Time* magazine, and the incident has since been recognized as an icon for "Sagebrush II." The year before, the Nevada Association of Counties endorsed the transfer of federal lands to the state, and Nye County further passed a resolution declaring that the state of Nevada owned public lands within the state and that the county had management authority over them.

This situation prompted a 1995 lawsuit from the federal government seeking to reaffirm that public lands belong to all Americans and are managed by the federal government and to protect federal employees from threats and intimidation. The legal action would continue beyond a 1996 court ruling until a 1997 settlement resolving all remaining issues, with the county agreeing to repeal its 1993 resolution.

On April 19, 1993, the FBI stormed the Waco, Texas, compound of the Branch Davidians, following a fifty-one-day siege. A fire broke out and consumed the building, killing seventy-six Branch Davidians, including twenty-five children, two pregnant women, and David Koresh, the leader of the group. *Mark Perlstein/Getty Images*

Supporters and members of the Jarbidge Shovel Brigade, led by Nevada Assemblyman John Carpenter, move "Liberty Rock" in 2000, opening the road to the Jarbidge Wilderness near Jarbidge, Nevada, a small town in northern Elko County. *Ross Endreson/The Daily Free Press*

But tempers hadn't cooled by the end of the decade. On July 4, 2000, Nevada Assemblyman John Carpenter led a group of 300 people to reopen a Forest Service road in the Jarbidge Wilderness near Jarbidge in Elko County; the Forest Service had elected to close it to protect a dwindling population of bull trout. Litigation over the Jarbidge Wilderness would continue until 2017, when a federal judge ruled against Carpenter's so-called shovel brigade, an event widely regarded as the end of Sagebrush II.

In 1993, a Nevada rancher named Cliven Bundy stopped paying his grazing fees after receiving an order to reduce the number of his cattle from the BLM, which was trying to protect the desert tortoise after it had been listed as a threatened species.[13] In 2014, the BLM attempted to round up Cliven Bundy's cattle, which at this point had been grazing illegally for twenty years, leading to a two-week armed standoff in Bunkerville, Nevada, that ended when the agency backed down and released the cattle.

On April 5, 2014, the Bureau of Land Management (BLM) attempted to round up approximately 900 cattle belonging to Bunkerville, Nevada, resident Cliven Bundy. Militia and property rights groups from around the country converged on the ranch, and an armed standoff ensued between the BLM and protestors. The BLM backed down, ending the roundup on April 12, but protestors lingered throughout April and into May, intent on defending Bundy and his rural way of life from the perceived heavy hand of the federal government. *Jim Urquhart/Reuters*

Two years later, Bundy's sons led a group that seized control of the Malheur National Wildlife Refuge in Harney County, Oregon, promoting the view that federal public lands should be turned over to state control. The occupation ended after a little more than a month, during which time one participant, Robert LaVoy Finicum, was killed during an attempted arrest.

The Forest Service revoked the grazing permits of Nevada rancher Wayne Hage in 1993 and, for allegedly violating the terms of those permits, impounded 100 of his cattle. Hage nonetheless continued to graze cattle on public land and took the Forest Service to court, claiming that he had water rights to the property. Hage's legal battle, continued by his son after his 2006 death, spanned twenty-five years and ultimately ended in favor of the government.

Back in Washington, D.C., members of Congress—ever sensitive to their prospects for biannual reelection—were taking note of the West's eruption over endangered species, property rights, and the federal government's authority. And judging by its record since 1988, Congress decided it wasn't going to deal with the ESA anymore.

CHAPTER 4

Babbitt's Jujitsu

IN 1990, A WOODPECKER BROUGHT OPERATIONS AT THE U.S. ARMY'S largest post to a virtual halt. Nestled among dense pine forests outside of Fayetteville, in the sandhills of central North Carolina, Fort Bragg (now Fort Liberty) was an increasingly important habitat for the endangered red-cockaded woodpecker. The FWS released a formal ESA opinion that found that training operations at Fort Bragg jeopardized the red-cockaded woodpecker's continued existence. Therefore, training operations on the post were limited to platoon-level exercises, and even platoons were limited to dismounted exercises. Night training was impacted as well. This situation ultimately led to the total failure of units at Fort Bragg at every level to meet their training objectives.[1]

At Fort Bragg, as at other installations across the Southeast, development pressure *outside* the fence line had driven wildlife, including endangered species, *inside* the fence line. Their protection under the ESA was interfering with military missions, placing the army into an impossible bind. On January 28, 1992, three civilian Department of Defense employees at Fort Benning, Georgia, were issued criminal indictments for destruction of red-cockaded woodpecker habitat and threatened with prison sentences of up to thirty-six years and fines of up to $650,000.[2]

Clearly, the ESA needed to be more flexible—in this case, to allow the army to do its job. However, it had to remain effective, to continue to act as Earth's emergency room. It was time for Congress to do its job and amend the law, as it had over the previous two decades since its creation. But now things were different.

Red-cockaded woodpeckers (*Leuconotopicus borealis*) building a nest. *Photograph by Doug Gochfeld*

At the time, a new generation of younger, more combative Republicans were winning election to Congress, riding the wave of the Sagebrush Rebellion, the Wise Use Movement, the County Movement, and a renewed focus on private property rights. These congressional Republicans began introducing bills with titles such as the Human Protection Act of 1991 (H.R. 3092) and the Balanced Economic and Environmental Priorities Act of 1991 (H.R. 4058), which would have amended the ESA to emphasize economic impacts in the listing process and/or to compensate landowners for regulatory takings under the ESA.[3] Needless to say, these bills were dead on arrival in the Democratic-controlled Congress. But they were markers that would frame the debate over the ESA throughout the 1990s and beyond. Republican members of Congress would not entertain improvements to the ESA because to do so would cross the increasingly powerful property rights lobby. And Democratic members of Congress would not admit that the ESA needed perfecting and cross vocal environmental activists.

In January 1992, George H. W. Bush's administration imposed a moratorium on new regulations, including ESA listings. The spotted owl debate was at its peak. And rather than deal with the suddenly controversial law in a presidential election year, Democrats allowed the ESA's authorization to lapse in 1992 (though it continued to function implicitly thanks to funding from Congress's parallel appropriation process).

Far from resolving tensions and allowing the ESA to move forward, the 1992 election made things more polarized. Democrat Bill Clinton won the presidential election and proceeded with his Northwest Forest Summit and subsequent Northwest Forest Plan, keeping the spotted owl issue front-page news. Several newly elected Republican members of both the House and the Senate were opposed to the ESA. Throughout the 103rd Congress (1993–1994), these members continued to introduce legislation focused on property rights, compensation for regulatory takings, and economic impacts of ESA listings.[4] The Democratic leadership of both chambers of Congress continued to defer action on the issue.

Congressional Republicans grew increasingly conservative, combative, and threatening toward President Clinton's agenda. In September 1994, Congressmen Newt Gingrich (R-GA) and Dick Armey (R-TX) would unveil the "Contract with America," propelling the party to the "Republican Revolution" that ended decades of Democratic control on Capitol Hill. Clinton had run on an anti-tax, pro-growth platform and, through two years, had largely failed to deliver either his ideas on energy policy or relief from the early 1990s economic recession. The dot-com boom in the stock market was still far in the future.

The 1994 midterm congressional elections brought a seismic shift to power on Capitol Hill, ending in an instant a long history of overwhelming Democratic Party control that stretched back to the days of President Franklin D. Roosevelt and the New Deal. The GOP gained fifty-four House and eight Senate seats, taking control of the House (for the first time since 1955) and the Senate as well. The new Republican majority would clash with the Clinton administration on nearly every issue imaginable.

NO SURPRISES

A dedicated conservationist and former governor of Arizona, Bruce Babbitt, President Clinton's secretary of the interior, understood the mindset of westerners. An alliance between extractive industries (mining, oil, and gas) and a Republican Party increasingly representative of private industry and property rights would present a formidable threat to the environment. The spotted owl controversy was the tip of the GOP spear.

Babbitt saw HCPs—which saved the San Bruno Mountain butterflies—as a potential solution to the problem of northern spotted owl management. But by the end of 1992, only fourteen HCPs had ever been developed, *including* the San Bruno Mountain HCP.[5] In practice, the HCP development process established by Congress in 1982 proved too complex, too unwieldy, and too uncertain for private companies to regard HCPs as a good investment. Few possessed the patience, resources, and history of collaboration and trust that had made the San Bruno Mountain HCP possible. Still, Babbitt actively lobbied timber companies in the region to adopt them.

In 1993, the Tacoma, Washington–based Murray Pacific Corporation agreed to an HCP covering 53,527 acres for the northern spotted owl.[6] Unfortunately, soon after the HCP was complete, disaster struck. Marbled murrelets were found on the property, and the company's spotted owl HCP and associated ITP provided no protection for taking murrelets.[7] The company's plans were thrown into doubt.

Babbitt and others, including Vice President Al Gore, lobbied Murray Pacific to expand its HCP to cover marbled murrelets. But given the inadequacy and failure of its spotted owl HCP, why should the company make the considerable investment of time and resources in *another* HCP that would be just as vulnerable to changing circumstances as the first?

Confronted by Murray Pacific's concerns, Secretary Babbitt and his team developed the concept of "no surprises" for HCPs. Initially promulgated as a policy in August 1994 and elevated to a regulation in 1998, "no surprises" states simply that "no additional land use restrictions or financial compensation" will be required of an ITP holder, "even if unforeseen circumstances arise after the permit is issued indicating that additional mitigation is needed."[8]

In other words, if a private landowner goes through the lengthy process of developing an HCP and securing an ITP, the U.S. government will honor that permit as a binding commitment on *itself* as well, and the permit holder can be confident that nothing more will be required of them for the species covered by the HCP. In 1995, Murray Pacific replaced its initial northern spotted owl HCP with a revised multispecies HCP with no surprises.[9]

Bruce Babbitt (born 1938) served as secretary of the interior for all eight years of Bill Clinton's presidency, following service as attorney general of Arizona (1975–1978) and governor of Arizona (1978–1987). A practicing attorney, Babbitt also served on the President's Commission on the Accident at Three Mile Island and made an unsuccessful run for president of the United States in 1988. As secretary, Babbitt skillfully navigated the treacherous political waters of the 1990s, leading a team that developed new regulatory flexibilities for the Endangered Species Act (ESA) and helped blunt the wave of anti-ESA sentiment in the American West and the Republican Party. A lifelong conservationist, Babbitt is particularly dedicated to America's public lands and their preservation and accessibility for recreation by all Americans. He has served on the boards of the World Wildlife Fund (U.S.) and the Amazon Conservation Association, and he remains active in biodiversity, wilderness, and public lands advocacy. *Gage Skidmore/Wikimedia Commons*

They were soon followed by others, such as Plum Creek Timber, and by the end of the Clinton administration, a total of 177 new HCPs had been approved, covering more than 26 million acres.[10] Thanks to "no surprises," today there are more than 1,400 HCPs encompassing more than 48 million acres.[11] This simple regulatory innovation—the basic idea that the federal government, too, must honor its commitments—made HCPs not just a legislative flexibility also but a regulatory flexibility and an opportunity that private companies were eager to avail themselves of. In the future, as Congress descended progressively deeper into partisanship and gridlock and effectively ceased to function, all new ESA flexibilities would be regulatory in origin.

"No surprises" was a solution not only to the practical problem of encouraging private investment in endangered species conservation but also to the political problem of increasing opposition to the ESA. Babbitt and his staff saw the coming Republican Revolution as a danger to the ESA, and with the "no surprises" policy, they struck first, co-opting the interests of industry and weakening public support for potentially damaging ESA amendments.[12]

In this way, Babbitt and his team hit on a strategy they would employ with great success throughout the rest of the Clinton administration, ultimately developing a number of regulatory flexibilities that are available under the ESA to this day. Don Barry, who helped implement the ESA in the 1970s and was Clinton's assistant secretary for fish, wildlife, and parks, colorfully refers to this as a campaign of "intellectual jujitsu."[13] To introduce this strategy to the American people and keep the political dialogue shifting in their favor, on March 6, 1995, the Department of the Interior released a white paper titled "Protecting America's Living Heritage: A Fair, Cooperative and Scientifically Sound Approach to Improving the Endangered Species Act."[14] The paper featured "Ten Principles for Federal Endangered Species Act Policy," which became known as Babbitt's "Ten-Point Plan."[15]

Although the Ten-Point Plan was aspirational and developed primarily to serve as political messaging to disarm the combative Republicans' repeated attacks on the ESA by removing issues they vigorously objected to, it captured the core tenet of deconflicting ESA enforcement against private parties to reduce real and perceived conflicts around the ESA.[16]

SOLVING FOR "SHOOT, SHOVEL, AND SHUT UP"

All of Secretary Babbitt's lieutenants were committed conservationists, and all worked well together, often meeting late into the night in the secretary's office to discuss the myriad issues and responsibilities managed by the sprawling Department of the Interior, which controls some 500 million acres, 67,000 employees, and a budget of $17 billion per year.

Babbitt and his brain trust also worked well with the Environmental Defense Fund's Michael Bean and Robert Bonnie, who met with the military and FWS to design an innovative conservation plan at Fort Bragg for the red-cockaded woodpecker. Bean was already a legend in ESA circles—an attorney by training, he was the nation's foremost expert on the ESA, a frequent witness at congressional hearings (where he testified with a professional demeanor), and well known in the Clinton administration—and the political leadership of the Department of the Interior was prepared to support any effort Bean was part of.[17] He was articulate and visionary, without his own private agenda, and thus trusted by both conservatives and liberals. Bonnie had a similar style and was hence a trusted adviser with an approachable, friendly, and inviting demeanor.

Bean, Bonnie, and their collaborators ultimately produced a specialized HCP that essentially established habitat for the woodpeckers on private land *off* the post to take pressure off Fort Bragg. But it incorporated an additional provision that stated that any landowner who voluntarily created red-cockaded woodpecker habitat would be protected from liability under the ESA should they later reverse their approach and modify that habitat. The only limit was that they could not reduce it below the "baseline" set when they enrolled in the program.[18]

Babbitt embraced this initiative, and the Clinton administration further developed it into a new conservation tool: a so-called safe harbor agreement (SHA). In 1999, the Fish and Wildlife Service and the National Marine Fisheries Service (collectively known as "the Services") implemented the new program through enhancement of survival permits under section 10(a)(1) of the ESA rather than ITPs under section 10(a)(2). This approach simplified the new tool and allowed the services to develop a new permitting procedure.[19]

SHAs were a critical innovation and a turning point for the ESA because they resolved the perverse incentives that led landowners to "shoot, shovel, and shut up" protected species. Prior to this time, the ESA had only penalized and regulated landowners. If a protected species were found on their property, at *best* they could secure an ITP and continue their ordinary activities with no further cost beyond the development and implementation of an HCP, which was itself a considerable cost. But if there were no protected species on their property, that would be better for them. And it would be truly insane for a landowner to attempt to create habitat and encourage protected species to colonize their property. And therein lay the perverse incentive: endangered species need more and better habitat in order to recover. But were landowners to create such habitat, they could only suffer. Hence, the incentive built into the ESA was for landowners to avoid creating habitat for endangered species at all costs.

This was a recognized phenomenon in the case of the red-cockaded woodpecker. Red-cockaded woodpeckers nest in cavities they excavate from mature pine trees, and they need trees of a certain diameter—and therefore age—in which to nest. Studies have found that landowners in red-cockaded woodpecker country would log their property frequently in order to ensure that no red-cockaded woodpecker habitat would develop.[20] The SHA tool resolved this perverse incentive: SHA permits allow landowners to return their property to its "baseline" conditions— before they took action to protect, improve, or provide habitat for endangered species—without negative repercussions, even if doing so results in take of listed species. Landowners further receive assurances that "additional conservation measures will not be required and additional land, water, or resources use restrictions will not be imposed should the covered species become more numerous as a result of the property owners' actions."[21]

The impact of the SHA tool was immediate and profound. Today, there are more than 100 SHAs in effect across the country, covering dozens of species and millions of acres of land.[22] For the red-cockaded woodpecker specifically, there are now eleven active SHAs covering 400 landowners and conserving 2.5 million acres of habitat.[23] These agreements allow landowners to take actions that create red-cockaded woodpecker habitat—such as the regular burning of forests; forest management for larger, mature trees; and the installation of artificial nesting cavities to accelerate colonization—without fear of negative consequences under the ESA. The species' population is growing, and delisting of the species may be possible as soon as 2030.[24]

SHAs were designed to entice private landowners to conserve threatened and endangered species. Because it is always easier to conserve a species before it reaches the point of listing under the ESA, a logical complement would be a similar regulatory tool for *candidate* species. The Clinton administration developed just such a tool in parallel to SHAs: candidate conservation agreements with assurances (CCAAs). In a CCAA, a landowner voluntarily agrees to implement conservation measures to benefit one or more covered candidate species. In exchange for their voluntary efforts on behalf of these species, the landowner receives regulatory assurances that additional conservation measures above what they have agreed to will not be required, even if the species become listed under the ESA, and that they may continue existing and agreed-on land uses that have the potential to incidentally cause take as long as the level of take is consistent with the level identified and agreed on in the CCAA.[25] Depending on circumstances, suitable conservation actions may include long-term management actions, such as prescribed fires; one-time construction of a habitat feature, such as breeding ponds; or the one-time removal of a threat, such as grazing.[26] In some cases, merely agreeing to continue present beneficial land management practices into the future may be sufficient.[27] Today, there are more than sixty approved CCAAs.[28]

The twin innovations of SHAs and CCAAs represented a major shift for private landowners and the ESA. Prior to their development, private landowners could receive only an ITP—and only after completing the arduous and expensive process of preparing an HCP. Enhancement of survival permits were reserved largely for people and organizations engaged in captive breeding of listed species, which may occasionally cause inadvertent take.[29] After the Clinton administration developed SHAs and CCAAs, enhancement of survival permits became available to private landowners.

Artificial nesting cavities have been key to red-cockaded woodpecker recovery. The species rebounded from a population low of 1,981 breeding groups in 1990 to 7,794 breeding groups in 2020 thanks to conservation measures such as this. *Courtesy of Raven Environmental Services, Inc.*

With HCPs, Congress made it possible for private landowners to touch the habitat of a protected species without necessarily triggering the prohibitions and enforcement of the ESA. With no surprises, the Clinton administration made HCPs a tool that private landowners felt more comfortable relying on. And with SHAs and CCAAs, the Clinton administration removed the perverse incentive for landowners to destroy habitat, meaning that it was no longer *harmful* to their self-interest for landowners to conserve threatened and endangered species. But there was still little in the way of *incentives* for landowners to conserve listed species. For landowners to be incentivized, to be better off because they actively conserved listed species, something more would be needed. And the most obvious incentive is funding, of which the ESA provides precious little to anyone and none to private landowners. Funding private lands conservation would instead fall to other programs, such as the federal Farm Bill, and to this day the need far outstrips the funding available.

REPUBLICAN ATTACKS ON THE ESA

Following the 1994 midterm elections, Republicans saw their ascension to the majority in Congress as a mandate to slash government spending and balance the federal budget. They aimed to reduce the scope of welfare and environmental laws—including the ESA—and other federal programs that were legacies of the liberal post–New Deal era of American government. Congressional Republicans ultimately overreached and achieved little of their agenda and were outmaneuvered by President Clinton, who vetoed many of their bills and co-opted issues such as welfare reform and a balanced budget.[30] Although the ESA was squarely in the GOP's sites, few of their legislative attacks gained traction.

The lone successful attack on the ESA occurred in 1995, when a rider on a must-pass Department of Defense supplemental appropriations bill imposed a moratorium on all ESA listings and critical habitat designations, lasting from April 10, 1995, until April 26, 1996.[31] But later in 1995, a Department of the Interior appropriations bill with much deeper cuts to the ESA was defeated due to a lack of Republican support. Instrumental to its defeat, ironically, was Speaker of the House Newt Gingrich. Gingrich was a well-known animal lover who famously started his political career at age eleven by petitioning his hometown of Harrisburg, Pennsylvania, to open a local zoo. His childhood heroes included visionary conservationists such as Raymond Ditmars and William Temple Hornaday, and throughout his political career, he made a point of visiting zoos wherever he traveled.[32]

No less an authority than Michael Bean himself wrote in 1999 that Gingrich "saved" the ESA from "an almost certain dismembering."[33] In 1995, Representative Richard Pombo (R-CA) chaired the special bipartisan Task Force on the Endangered Species Act that conducted oversight hearings and considered possible amendments to the ESA. In testimony before the task force, Gingrich opposed any changes to the ESA that prioritized property rights at the expense of conservation. As he put it, "On the one front, you have legitimate property rights and legitimate economic interests. On the other front, you have a need for a level of biodiversity and a level of concern for the biological system that is also, I think, paramount. And the question is, how can we work together?"[34] Gingrich further cautioned,

> If we of this generation destroy the resources from which our children would otherwise derive their livelihood, we reduce the capacity of our land to support a population, and so either degrade the standard of living or deprive the coming generations of their right to life on this continent.[35]

Gingrich had previously cosponsored pro-ESA legislation, and he further used his position as Speaker to derail Republican efforts to gut the ESA, much to the irritation of the House Republican Caucus. His lack of support was critical in preventing more successful legislative attacks on the ESA.[36]

Although the Republicans remained in control of both the House and the Senate after the 1996 elections, the party had been bloodied by its own bellicose implementation of the Contract with America and chastised by the reelection of President Clinton. Emerging in this somewhat softened state was a bipartisan attempt to reauthorize and amend the ESA, led by western senators Dirk Kempthorne (R-ID), Max Baucus (D-MT), and Harry Reid (D-NV). They were joined by the long-serving John Chafee (R-RI). Their legislation would have provided political victories to both the left and the right; it addressed data quality, recovery planning, delisting, public input, state conservation agreements for candidate species, funding for individual conservation efforts, federal interagency consultations, ITPs, HCPs, and SHAs.[37]

Newt Gingrich (born 1943) served as the fiftieth Speaker of the United States House of Representatives (1995–1999). A Republican from Georgia, Gingrich led the pugnacious Republican House majority in its many confrontations with Democratic President Bill Clinton. But as a lifelong animal lover, Gingrich opposed Republican efforts to undermine the Endangered Species Act and personally stopped a number of Republican attacks on the law during his speakership. Throughout his career, Gingrich was known for visiting zoos on the campaign trail and when traveling. Here, he holds a red-tailed hawk during a visit with Columbus, Ohio, Zoo officials in his Capitol Hill office, June 27, 1995. *Joe Marquette/Associated Press*

It went through the hearing and amendment processes and was expected to come to the Senate floor, pass, and proceed to the House. Tragically, in a harbinger of things to come, last-minute maneuvering behind the scenes by both environmentalists on the left (who wanted a more protective bill) and conservative Republican members of Congress (who wanted a less protective bill) killed the legislation. Although many environmental groups had participated in development of the legislation, they failed to support the final product, and many of their grassroots members were firmly opposed to it.

Congress's most serious (though unsuccessful) threat to the ESA would materialize nearly a decade later—a resurrection of the western strain of pugnaciousness that characterized the Sagebrush Rebellion. The Threatened and Endangered Species Recovery Act of 2005 (TESRA) was sponsored by Republican congressman Richard Pombo, a rough-and-tumble, hard-core private property rights farmer from California's Central Valley. Five feet, six inches tall with a goatee and mustache, stocky and of Portuguese descent, he was feisty, sharp, and quick in temperament, always fighting to maintain his position. He had been a vocal opponent of the ESA since first taking office in 1993. TESRA was actually Pombo's twelfth ESA bill in eight years, and as chairman of the House Resources Committee, a position he had assumed in 2003, he was well positioned to see it passed into law.[38]

TESRA was an unequivocally bad bill, one that would have effectively gutted the ESA's purpose. It incorporated some features of the earlier bipartisan Kempthorne/Chafee/Baucus/Reid bill. But it went much further, injecting economic, national security, and other factors into the listing process; eliminating critical habitat entirely; and imposing compensation requirements on the federal government in which conservation actions would be required by the ESA.[39] These changes would have made the ESA program far more expensive and unwieldy, made it easy to choose not to list a species on the basis of economic concerns, and eliminated essential protections for wildlife habitat.

TESRA passed the House on a bipartisan basis, with support from thirty-six Democrats canceling out opposition from thirty-four Republicans, for a final tally of 229–193. Environmental groups were vociferous in their opposition to TESRA. National Wildlife Federation attorney John Kostyack led a public relations blitz against the legislation, calling it "an affront to American values" and "poison for America's wildlife."[40] The total elimination of critical habitat and the staggering expense of the proposed compensation payments to impacted landowners were focal points for criticism.

In press materials, environmental groups mockingly referred to TESRA as "Pombo's extinction act." They and their allies ultimately succeeded at killing the bill in the Senate and in defeating Pombo in his 2006 reelection bid—highly unusual for a seven-term, fourteen-year member of Congress! Environmental groups such as Defenders of Wildlife, the Sierra Club, the League of Conservation Voters, and the Humane Society of the United States were furious about TESRA and Pombo's other anti-environment, pro-industry positions and spent between $1.5 million and $2 million to defeat him.[41] Pombo became a proxy for the environmental community to demonstrate that they could defeat an entrenched incumbent Republican.

Richard Pombo (born 1961) represented California's 11th Congressional District from 1993 to 2007. A radical property rights advocate, Pombo sat on and later chaired the House Committee on Resources, from which position he attacked numerous environmental laws, most notably the Endangered Species Act. He lost his reelection campaign in 2006—highly unusual for a congressman in his seventh term—in large part due to his 2005 attempt to overhaul the ESA. *MCT/Getty Images*

Unfortunately, the lesson that members of Congress took from TESRA was not that *bad* ESA legislation should be avoided but that *all* ESA legislation should be avoided. Dirk Kempthorne, who as a senator had been one of the sponsors of the bipartisan 1997 legislation that was defeated by extremists on both sides of the aisle (including, ironically, Richard Pombo), became George W. Bush's second secretary of the interior in 2006. From his position in the Secretary's Office at Interior Headquarters on C Street, Kempthorne was interested in reviving attempts at bipartisan ESA legislation, where executive agencies could push Congress to take action, following in the tradition of officials such as Nathaniel P. Reed, the Nixon administration's assistant secretary of the interior for fish, wildlife, and parks. Following the failure of TESRA, Kempthorne abandoned the idea.[42]

While the ESA escaped the Republican Revolution largely unscathed in substantive terms, it took a walloping in the realm of political rhetoric. The introduction of partisan amendments and appropriations riders designed to undermine and weaken the ESA became common, and their use has only increased in the years since. A pair of complementary studies carried out by the author and by the left-wing Center for Biological Diversity (CBD) have identified more than 600 such attacks in the period from 1995 through March 2021.[43] This situation in turn has driven campaigning, fund-raising efforts, and partisan messaging on both the left and the right.

This partisan gridlock and political brinksmanship—in stark contrast to the previous era of bipartisanship that produced the ESA—has prevented Congress from amending it since 1988. Legislative action by democratically elected representatives has failed Earth's emergency room. Therefore, regulatory fixes like the kind that Secretary Babbitt made, originating from within the executive branch, appear to be the only way to keep the ESA viable and relevant going forward.

CHAPTER 5

The Far Left's Listing Wars

NOT ALL ATTACKS ON THE ESA ORIGINATE ON THE RIGHT. "WE'RE crazy to sit in trees when there's this incredible law where we can make people do whatever we want,"[1] says Massachusetts-born and -educated Kieran Suckling, referring to the ESA, which his organization lucratively weaponizes against industry and even government. Suckling, the political polar opposite of Richard Pombo, embodies the radically left-wing "deep ecology" mindset: property rights and states' rights are not legitimate; animal and plant rights are. Slender, bearded, and balding, Suckling was described in a profile in *The New Yorker* in 1999 as a "trickster, philosopher, publicity hound, master strategist, and unapologetic pain in the ass."[2] He is an uncompromising idealist, influenced by thinkers such as Irish revolutionary Wolfe Tone, Saint Francis of Assisi, Martin Heidegger, and Jacques Derrida.

No matter the resulting collateral economic and social damage, deep ecologists believe that every species has intrinsic worth separate and apart from its financial value to humans and must be protected. Theirs is a religion of disdain for humankind's supremacy in the world order; it insists that modern society bend to the needs of wilderness and the natural world and live within its constraints. It is an intellectual movement and path to self-realization, using the primacy of nature and the wilderness for personal growth and a moral compass for action. Norwegian eco-philosopher Arne Naess first used the term "deep ecology" in 1973, advocating environmental harmony and systemic equilibrium.[3]

To achieve this utopian vision, Suckling, along with cofounders Robin Silver and Peter Galvin, created the CBD in 1989. It is one of the most dynamic, aggressive, and influential environmental groups in America today, although it has remained small, with a staff of approximately 175 nationwide and an annual budget of approximately $25 million. Its main approach has been to utilize the ESA, especially the petition process and accompanying judicial remedies, to force the protection of species (and thus their habitats) to change development and human land-use patterns.

To deep ecologists, applying the ESA to the spotted owl was a convenient pretext for also saving old-growth forests in the Northwest: all species of flora and fauna are the moral equivalent to animals, including humans. Earth and its diverse species are a holistic biosphere, not to be commoditized for human consumption and profit. Our anthropocentric government must be neutered, they believe, and the ESA is the perfect tool to accomplish that purpose. This approach, leveraging the ESA to intentionally create conflict, is not popular, and Suckling and his colleagues are widely disparaged across the political spectrum but also grudgingly respected for the quality of their legal and research teams and their effectiveness.[4]

Suckling's CBD employs nonviolent tactics, but its philosophical underpinnings are shared by the most notorious environmental groups—the Animal Liberation Front (ALF) and Suckling's previous organization, the Earth Liberation Front, both listed by the FBI as ecoterrorists for their use of incendiary devices, Molotov cocktails, firebombs and pipe bombs, burglary and arson, murder and attempted murder, ramming and sinking whaling ships, seizure and destruction of drift nets, interdiction of Canadian seal hunters, and the release of laboratory animals. "Arson, property destruction, burglary, and theft are acceptable crimes when used for the animal cause. . . . Damaging the enemy financially is fair game," said Alex Pacheco of People for the Ethical Treatment of Animals, which reportedly provides leadership for ALF.[5]

ECO DEPTH GAUGE

How deep is your ecology?
Take a sounding.

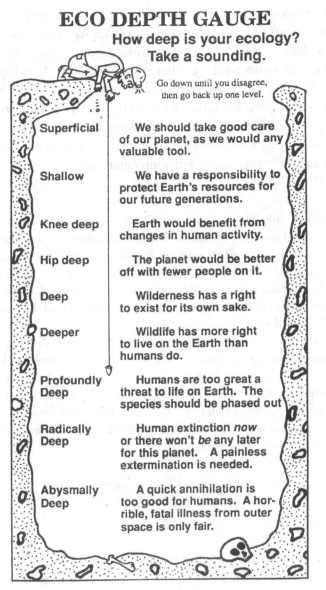

Go down until you disagree,
then go back up one level.

Superficial — We should take good care of our planet, as we would any valuable tool.

Shallow — We have a responsibility to protect Earth's resources for our future generations.

Knee deep — Earth would benefit from changes in human activity.

Hip deep — The planet would be better off with fewer people on it.

Deep — Wilderness has a right to exist for its own sake.

Deeper — Wildlife has more right to live on the Earth than humans do.

Profoundly Deep — Humans are too great a threat to life on Earth. The species should be phased out

Radically Deep — Human extinction *now* or there won't *be* any later for this planet. A painless extermination is needed.

Abysmally Deep — A quick annihilation is too good for humans. A horrible, fatal illness from outer space is only fair.

"Deep ecology," a term coined by Norwegian eco-philosopher Arne Naess in 1973, is a philosophy that holds that all life has inherent worth, as does Earth and its inanimate features. Humans are equal with all species and have no moral, ethical, or legal right to dominate or destroy Earth, the environment, or other species. *Produced by Les U. Knight*

In their own way, Suckling's bureaucratic attacks jeopardize Earth's emergency room as much as Pombo's TESRA legislation: Suckling and his allies alienate reasonable conservationists, undermine the credibility of government actors working diligently to save endangered species, and divert public resources away from conservation efforts and toward legal protection. Using the ESA, the richly compensated Suckling and his colleagues at the CBD have found that lawsuits are more effective than violence at subverting Congress's intent and easier to market to well-heeled donors eager to contribute to environmental causes.

Suckling explains,

> We use lawsuits to help shift the balance of power from industry and government agencies, toward protecting endangered species. That plays out on many levels. At its simplest, by obtaining an injunction to shut down logging or prevent the filling of a dam, the power shifts to our hands. The Forest Service needs our agreement to get back to work, and we are in the position of being able to powerfully negotiate the terms of releasing the injunction.
>
> New injunctions, new species listings and new bad press take a terrible toll on agency morale. When we stop the same timber sale three or four times running, the timber planners want to tear their hair out. They feel like their careers are being mocked and destroyed—and they are. So they become much more willing to play by our rules and at least get something done. Psychological warfare is a very underappreciated aspect of environmental campaigning.[6]

The costs of Suckling's attacks are not merely psychological. Listing, recovery, and eventually—hopefully—delisting species is a monumental undertaking.

MONKEY WRENCHING AND MEGAPETITIONS

When a species is "listed" under the ESA, there are real consequences. It is not just a symbolic gesture or a means of alerting the public to the need to conserve a species. On the contrary, the ESA lays out restrictions against harming a listed species or its critical habitat, and these are enforced by both civil and criminal penalties. It also creates affirmative conservation obligations, primarily for federal agencies. Finally, it triggers a vast array of conservation programs and opportunities, which can marshal significant funding, attention, and third-party involvement. Listing, in other words, is a serious business.

Kieran Suckling, executive director of the Center for Biological Diversity (CBD), and a man dressed as the CBD's mascot Frostpaw the Polar Bear in a street protest near the Charging "Wall Street" Bull statue in New York's financial district, September 22, 2014. *Steve Rhodes/Corbis*

When "Sagebrush Rebel" Ronald Reagan took office in 1981, listings ground to a halt under his first secretary of the interior, the pro-industry, anti-environment James Watt. Watt wasn't acting alone: the Reagan Revolution set about shrinking all government regulatory agencies established during the Green Revolution, including the EPA. Congress rose to the occasion and passed a 1982 amendment to the ESA requiring the FWS and National Marine Fisheries Service to respond promptly to any petition, whether to list or delist a species or to designate critical habitat, within ninety days. Within that period, the FWS must determine whether the petition presents enough information about a species to potentially necessitate action. If it does, then whether the requested action is actually warranted must be determined within twelve months from the original petition date. The amendment also expanded the ESA's citizen suit provision to allow lawsuits to compel federal agencies to meet these deadlines.

Ironically, the 1982 amendment protecting ESA operations from right-wing inertia opened the door to left-wing attacks like those organized by Kieran Suckling. Suckling and his colleagues at the CBD have used that amendment to make sure that the FWS can't do its job and actually screen petitions for their validity. The CBD intentionally floods the FWS with petitions and lawsuits to prevent it from operating with discernment. "Our modus operandi is to take [opponents] by storm," says Suckling. "We don't let the industry or agencies know what we're doing because they'll try and stop us. But once we file a petition or lawsuit, they can't respond quickly enough. Then we file another. It's like boxing. We hit them once and before they have a chance to recover we hit them again, and we keep hammering away until they fall down."[7]

As early as 1992, the FWS faced deadline litigation for hundreds of candidate species, leading to a massive negotiated settlement that set the agency's agenda for 1992–1996.[8] This is a far cry from what Congress had in mind in 1982 when it amended the ESA to make sure each petition would be scrutinized carefully.

In the 2000s, environmental groups (including the CBD) began filing "megapetitions," seeking listing of hundreds of species at a time. These megapetitions created a situation so unworkable that the FWS was forced to consolidate twelve different cases into a single "multidistrict litigation" in order to reach a global settlement with the CBD and its allies. Ultimately, it reached settlement agreements that dictated the agency's listing program from 2011 to 2016, including listing or critical habitat determinations for 1,030 species, subspecies, and populations and 1,559 judicially enforceable deadlines.[9]

Edward Abbey's 1975 novel *The Monkey Wrench Gang* lionized the acts of mechanical sabotage that four committed environmental activists perpetrated against strip miners, road and dam builders, and real estate developers. While some members of the deep ecology movement resorted to violence and property destruction in real life, most of the monkey wrenching perpetrated by groups such as the Center for Biological Diversity has been legal and bureaucratic in nature. *Edward Abbey/*The Monkey Wrench Gang. *Copyright by Ken Sanders and Dream Garden Press. All rights reserved.*

Not only were these megapetitions preventing serious scrutiny of endangered species, but they were also draining the agency's resources. In 2011, the allocation for listing and critical habitat determination was $20.9 million, and the agency spent at least $15.8 million—75 percent of its budget—taking substantive actions required by court orders or settlement agreements resulting from litigation.[10]

In 2016, the Obama administration adopted new regulations to ban "megapetitions" by limiting each petition to a single species,[11] but the agency could still be overwhelmed if environmental groups simply filed more petitions for individual species. Experience has shown that when the FWS faces an unmanageable volume of petitions, other work suffers. Even the petitioned species themselves may suffer, as the FWS—constrained by the budget appropriated by Congress—can only make the required ninety-day findings and then make the required twelve-month findings. In the meantime, new petitions, including petitions for potentially more direly imperiled species, pile up unaddressed.

The overwhelming majority of listings are not initiated by the government but are granted in response to petitions from organizations and the public. Most lawsuits by petitioners result in the Services losing (or, more often, settling the case and agreeing to complete the listing determination), giving courts and litigants substantial influence in the listing process.

When the CBD and organizations like it monkey wrench the legal system and the federal government by submitting endless ESA petitions and filing round-robin serial lawsuits, their publicists cynically spin their bureaucratic sabotage as evidence of the FWS's negligence in missing statutory deadlines. Donations from satisfied funders stream in, and these far-left organizations are then amply positioned to strike again. The cycle is fodder for a news media content to report that the FWS and the ESA have violated the public trust over and over again and for a public ready to believe negative press about its government.

The truth is that there is not nearly enough public money allocated to the FWS and the National Marine Fisheries Service to enable them to list all of the species that qualify for listing or even to address all of the petitions received, to say nothing of the critical habitat designations, recovery plans, five-year evaluations, consultations, permit applications, delisting, and post-delisting monitoring that ought to follow. Moreover, such money as is available is subject to reallocation to deal with pressing administrative needs, which may not be the same as pressing *biological* or *ecological* needs.[12]

As of December 2023, 1,668 domestic species of plants and animals were listed under the ESA. An additional 698 foreign species were listed as well. One hundred twenty-six species had been delisted—seventy-three due to recovery, thirty-two due to extinction, and twenty-one because they were listed in error.[13]

The "Agony" of Critical Habitat Designations

The ESA defines "critical habitat" as "specific areas . . . essential to the conservation of the species." Critical habitat designations are complicated, riddled with exceptions, and controversial even though most critical habitat is on public lands. These areas may be presently occupied or unoccupied but of potential value in the future, and the procedures for identifying each in turn are controversial. Daniel M. Ashe, FWS director during the Obama administration, observed,

> Whenever we designate critical habitat, it is a firestorm and it doesn't really do very much. It's like juice that's not worth the squeeze. There's a very limited range of circumstance where critical habitat really adds value to conservation, but it always causes disproportionate political controversy because you're publishing a map, and people, when they see a map and their property is inside of the map, they're entitled to assume significance, but it has little significance. So that's probably the most difficult part of managing the law, and it's very difficult, and I think for the future, we need to figure out a way to deal with that.[14]

As of December 2023, designated critical habitat covered approximately 107.5 million distinct acres as well as 36,000 distinct stream miles. A further 3.6 million acres and 200 stream miles of critical habitat were proposed pending final designation.[15] This is almost 168,000 square miles, or 4.4 percent of the entire United States—not enough, in Kieran Suckling's worldview.

For many years, the Services deprioritized critical habitat designation until a series of successful environmentalist lawsuits in the 1990s forced them to make prompt critical habitat designation a regular part of listing species. The CBD petitioned to list the Mexican spotted owl in 1989 and obtained internal FWS memos showing that the Bush administration had no intention of designating critical habitat for it.[16] ("Critical habitat has turned out to be an agony!" distinguished legal scholar Oliver Houck observed in 1993.[17])

Worse still, after winning a summary judgment against the FWS, the CBD also proved that the U.S. Forest Service—in approving a timber sale—failed to consult with the FWS to determine whether the owl's critical habitat would be disturbed. U.S. District Judge Carl Muecke made the sale contingent on a biological opinion—which the Forest Service produced in superficial form at the close of business on a Friday. Logging in the owl's critical habitat began the following Tuesday.

Judge Muecke was outraged and threatened to hold the regional forester and/or chief of the U.S. Forest Service in contempt. The injunction order was then honored but only after 60,000 board feet of timber were felled.[18] The federal government's interests were clearly anthropocentric—putting loggers before owls and putting itself on war footing with the CBD ever since.

The ESA commands all federal agencies to ensure "that any action authorized, funded, or carried out by such agency . . . is not likely to jeopardize the continued existence of any endangered species or threatened species or result in the destruction or adverse modification of [its critical] habitat." This objective can affect businesses, industry, state and local governments, infrastructure projects, and any number of other activities that are *not* carried out by a federal agency if private actors lease federal lands, their activities require federal agency permits, or they are otherwise subject to a "federal nexus," which is a connection to a federal agency through an agricultural or other government program, including federal crop subsidies, crop insurance, flood insurance, federal loans, disaster relief payments, permits under federal antipollution laws, or enrollment in federal conservation programs under the Farm Bill or other laws.

So if a private landowner has a "federal nexus," they can be subject to critical habitat designations—which they will usually fight tooth and nail. Edward Poitevent, for example, owned 1,544 acres of land in Louisiana that the FWS included in a 6,477-acre critical habitat designation for the dusky gopher frog. His land, which he leased to the Weyerhaeuser timber company, contained the kind of ephemeral ponds that dusky gopher frogs use to reproduce. But the ponds on the Poitevent property were in unoccupied habitat—in mature forests instead of the young forests required by the frog. So the land couldn't even support the frog unless it was substantially modified.

The landowners took their case all the way to the U.S. Supreme Court and won in 2018—or at least won an order for the FWS to reconsider its decision because the affected private land was not usable *habitat* for the frog, let alone "critical" habitat.[19] As the plaintiff put it, "To make it suitable you'd have to rip up every tree, . . . replant all of it with the right tree, make sure the ponds are still there, and make sure you burn it every year."[20] The plaintiff won not just a moral victory but also a very real financial one, as the FWS estimated that the critical habitat designation could lead to $34 million in economic losses.

The methodology for determining a critical habitat's economic impact is also controversial. The Services evaluate the economic impact of critical habitat using an "incremental" approach, which discounts any economic impact that is coextensive with the economic impacts of listing the species in the first place. As a result, many critical habitat determinations assert that there are "zero" economic impacts of the designation, much to the frustration of not only private landowners but also ranching, mining, oil and gas companies, and others that lease public lands as part of their businesses.

Areas may be excluded from critical habitat if "the benefits of such exclusion outweigh the benefits of specifying such area as part of the critical habitat." Hence, how these impacts are evaluated matters. Critical habitat designation may be forgone if it can't be determined or if its designation is "not prudent" (for example, if doing so would publicly identify the location of a rare species or dangerously delay the listing of one critically imperiled).

The dusky gopher frog (*Rana sevosa*) was listed as an endangered species in 2001. The wild population of the frog is approximately 135, spread out across six ponds, with only a single pond hosting a breeding population, making the species extremely vulnerable to extinction in the event of disease, natural disaster, or another abrupt change to its habitat. The U.S. Fish and Wildlife Service designated critical habitat for the frog in 2012, mostly on public land in Mississippi, but it also designated a parcel of private land in Louisiana, a state where the frog had not been sighted since 1965. The Louisiana parcel, in St. Tammany Parish, further lacked the necessary features for frogs to inhabit it. This situation led to the Supreme Court case *Weyerhaeuser Co. v. U.S. Fish & Wildlife Service*, in which the Court held unanimously that only habitat suitable for the species could be designated as "critical habitat." *Angela Dedrickson/U.S. Fish and Wildlife Service*

THE FRECKLEBELLY MADTOM HOLDUP

In 2010, Suckling's CBD filed a petition for the FWS to list 404 aquatic, riparian, and wetland species in the southeastern United States. Among them was the frecklebelly madtom (*Noturus munitus*), a small, nocturnal catfish that depends on cool, clear, moving water undisturbed by sedimentation. The species was once broadly distributed across the Southeast, but as its population dwindled, it became increasingly rare. It is threatened by habitat destruction and degradation caused by agriculture and development, resulting in poor water quality; habitat degradation from channelization, dams, and impoundments; and climate change.

On March 2, 2023, the FWS listed as threatened a distinct population in the Upper Coosa River basin in northern Georgia. At the same time, it also designated critical habitat and soon thereafter (on May 25, 2023) published an interim recovery outline to guide conservation while a full recovery plan is developed.[21] At first glance, the frecklebelly madtom appears to be an exemplar of a smooth listing process, with the listing and critical habitat designation for the species coming together and recovery planning soon after. But in the more than a decade that elapsed between its listing being petitioned and completed, the species was affected by efforts that represent both the best and the worst of the ESA.

The CBD's megapetition that included the frecklebelly madtom along with 403 other species was the second-largest ESA petition ever filed (the largest was for 475 species, also in 2010 by WildEarth Guardians, which, like the CBD, has long been engaged in a saturation petition-and-sue campaign). The FWS was forced to consolidate twelve different cases into a single "multidistrict litigation" in order to reach a settlement with the CBD and WildEarth Guardians. The workload imposed by these settlements forced action on the madtom to be delayed, and the listing workplan that was developed following the multidistrict litigation called for a twelve-month finding and proposed rule to be delayed until 2020. The FWS duly issued its proposed listing rule on November 19, 2020, and then finalized its decision on March 2, 2023.

Despite their radical predatory methods and intentions, the aggressive tactics and crushing workload created by the CBD and the like have a silver lining: forcing the FWS and its partners to innovate. This has resulted in, for example, the FWS's listing workplan, which has enabled the FWS and numerous partners to launch innovative and effective conservation programs benefiting countless species, including the frecklebelly madtom.

The frecklebelly madtom (*Noturus munitus*) is a small catfish native to the southeastern United States. Petitioned for listing in 2010, in 2023 the U.S. Fish and
Wildlife Service listed as threatened a distinct population segment in the Upper
Coosa River basin in northern Georgia. The workload imposed on the FWS by
settlements with the Center for Biological Diversity and WildEarth Guardians forced
action on the madtom to be delayed. *Brett Albanese/Georgia Ecological Services
Field Offices*

While the madtom's listing decision was pending between 2010 and
2020, extraordinary conservation efforts arose across the Southeast. The
madtom was included as a species of "high" concern in the Southeast
Conservation Adaptation Strategy, a collaborative effort initiated in
2011 by the fifteen states of the Southeast and the federal government
with the goal of promoting conservation throughout the Southeast and
the Caribbean.[22]

The madtom is protected by state law in four states (Alabama,
Georgia, Mississippi, and Tennessee). It has also benefited from captive

breeding and propagation, land management practices to minimize disturbances to rivers (such as soil erosion and nutrient runoff), and federal financial assistance to private landowners (under the Farm Bill) for these activities.[23] Altogether, conservation efforts such as these in the Southeast have resulted in more than 250 successes. These include various positive actions under the ESA, such as withdrawn petitions, not substantial ninety-day findings, not warranted twelve-month determinations, species downlisted, and species delisted. The number of these "wildlife wins" is increasing each year.

When the FWS eventually evaluated the frecklebelly madtom in 2020, it identified six different populations of the species but ultimately listed only one of them. Because of voluntary conservation efforts undertaken to benefit the species throughout the decade, none of the other five populations warranted listing.

This is the potential of voluntary, collaborative prelisting conservation, and thanks to it, the frecklebelly madtom was listed in only one small population. Thanks to it, too, the species has strong prospects for conservation, recovery, and delisting, just as the creators of the ESA intended.

THE POLITICS OF RECOVERY AND DELISTING

Congress intended for listing and delisting to be symmetrical processes driven solely by science and the application of identical listing factors laid out in the ESA. In practice, listing and delisting can be both political and policy driven, with uncertain outcomes. And it can be easier to list species than to delist them because the ESA is a precautionary law. As the Supreme Court recognized in 1978 in the landmark snail darter case, *TVA v. Hill*, Congress said as much in 1973:

> From the most narrow possible point of view, it is in the best interests of mankind to minimize the losses of genetic variations. The reason is simple: they are potential resources. They are keys to puzzles which we cannot solve, and may provide answers to questions which we have not yet learned to ask.
>
> Who knows, or can say, what potential cures for cancer or other scourges, present or future, may lie locked up in the structures of plants

which may yet be undiscovered, much less analyzed? More to the point, who is prepared to risk . . . those potential cures by eliminating those plants for all time? Sheer self-interest impels us to be cautious.

The institutionalization of that caution lies at the heart of [the ESA].[24]

Legal scholars call this the precautionary principle, a broad statement that when dealing with scientific uncertainties with significant potential downsides (such as the extinction of an ecologically significant species), humans should seek first to do no harm.[25] Applying the precautionary principle to the situation of delisting a species under the ESA, the logical conclusion is that if we are uncertain what will happen to a species, the safer course is simply to keep it listed. In some cases, this uncertainty can lead to species remaining listed when they have recovered sufficiently to be delisted.

This was not the case for the snail darter, which was delisted in 2022. The petition to delist it came from none other than Zygmunt Plater, the crusading law professor whose efforts to save the species back in the 1970s had led directly to *TVA v. Hill*. Plater teamed up with retired FWS and U.S. Geological Survey fisheries biologist Jim Williams and the CBD, who trumpeted the delisting in a press release: "Imperilment comes all too quickly, but recovery takes time, and the snail darter has now met the goals in its recovery plan and no longer needs protection."[26] Ironically, even this ESA success story was part of the CBD's ongoing campaign to control the FWS. By filing a delisting petition in 2019, the CBD and the other petitioners were able to influence the delisting process and push the FWS to link delisting to continuation of specific dam management practices by the TVA, even after management of the species was theoretically returned to the state.[27]

Like listing, delisting can be initiated through the ESA's petition process in the same manner and with the same information as petitions for listing. Delisting petitions are governed by the same updated regulations adopted in 2016.[28] On receipt by the FWS or the National Marine Fisheries Service, they are subject to the same statutory deadlines and evaluations laid out in the ESA.

The other way species ready for delisting are identified is internally, most typically through species status assessments that the Services prepare as part of their five-year reviews of listed species. These five-year reviews have been part of the ESA since 1978, when they were added by amendment at the urging of Representative Don Young (R-AK). The purpose of five-year reviews is to determine whether "factual evidence indicates a change in status of any species." They were proposed by Young in response to the situation of the gray wolf in Minnesota, which was reclassified from "endangered" to "threatened" only after three and a half years of intensive lobbying.[29] Thanks to five-year reviews, the Services possess an internal mechanism to know when species need to be downlisted or delisted (and very occasionally uplisted) without any outside petition being filed.

According to FWS databases, there are fifty-nine currently listed species (or distinct population segments thereof) that have been downlisted from endangered to threatened at some point in their history. An additional nine species have been downlisted and subsequently delisted.[30] In contrast, just thirteen threatened species (or distinct population segments thereof) have had to be uplisted to endangered.

Delisting has accelerated since the Obama administration. Sixteen species were delisted in 2016 alone as part of a push at the end of the Obama administration to delist as many species as possible. In fact, the Obama administration delisted more species than all previous administrations combined: twenty-eight in total. (The delisting of a twenty-ninth species, the northern Rocky Mountain gray wolf, was subject to a legal challenge and not finalized until shortly after Obama left office.) Dan Ashe, the director of the FWS for most of the Obama administration, attributed "successes in recovering and delisting species" to "the Endangered Species Act's effectiveness and the diverse collaborations it inspires" and credited "partnerships developed and maintained by the Service [that] have sustained years of recovery efforts for a myriad of species."[31]

This relatively recent focus on delisting is not a repudiation of the precautionary principle but a recalibration in recognition of the fact that Congress intended species to be recovered and delisted, not to languish on the endangered species list forever. Species can be successfully delisted even when recovery is a long process, even when they remain reliant on conservation, and even when litigation must be overcome. But it must be acknowledged that much of our progress in this area has come in just the past decade and a half. Those who are skeptical of the effectiveness of the ESA—and perhaps suspicious of it because of how it impacts them, their states, and their livelihoods—must be acknowledged as well. Although the future of recovery is bright, its mixed record is undeniably a source of controversy that affects the ESA to this day.

RECOVERY PLANNING AND MANAGING DELISTED SPECIES

A central species in the evolution of recovery planning was the iconic peregrine falcon (*Falco peregrinus*). Following World War II, the widespread use of DDT and other pesticides caused the American peregrine falcon population to decline. This was because DDT reduces the calcium content of eggshells, making eggs thin shelled and prone to breaking. This effect was particularly pronounced in predatory birds such as the falcon, California condor, and most famously the bald eagle, in all of which environmental pollutants such as organochlorines become concentrated in a process known as bioaccumulation. As a result of growing public awareness and litigation stemming from Rachel Carson's 1962 book *Silent Spring*, DDT was banned in 1972.

The peregrine falcon was listed in 1970 under the predecessor 1966 and 1969 ESA laws. During the initial debate on the ESA in 1972–1973, the peregrine falcon was cited as an example of a charismatic species that Congress wished to save. But the importance of recovery *plans* was not self-evident in 1973, and the original ESA did not mention them at all. In 1978, when Congress amended the ESA to require recovery plans, witnesses from the Peregrine Fund and related conservation groups shared their experiences with peregrine falcon conservation and urged more centralization of the recovery planning process.

Recovery plans incorporate both objective, measurable criteria for recovery and site-specific management actions to achieve recovery. They further include intermediate steps toward recovery and estimates of the time and cost to carry out the recovery plan. Commentators have described these requirements as "minimal," and recovery plans developed over the years have been inconsistent in quality.[32] In developing recovery plans, the Services are supposed to prioritize those species most likely to benefit, and

Congress has specifically highlighted the need to resolve conflicts with construction, development projects, and other forms of economic activity.

Four separate recovery plans were in development for the peregrine falcon in different regions, which in the eyes of the expert witnesses entailed "unnecessary duplication of effort," whereas a single recovery plan would have been preferable.[33] Despite their inefficiency, the four recovery plans were duly completed and implemented between 1979 and 1984, and both public and private recovery efforts continued. The species was ultimately delisted in 1999, even though not every goal in every recovery plan had been met—in some areas, it was clear that the recovery plans were too conservative.

Recovery plans are not regulatory documents, and while they inform delisting decisions, they do not control them. This means both that a species may be delisted without necessarily meeting all recovery plan objectives and that meeting all recovery plan objectives will not automatically result in delisting. However, recovery plans often represent the best available science regarding a species and so inform other regulatory processes under the ESA.[34] Courts also may rely on them as a source of scientific information about listed species.[35]

Even after decades of improvement, recovery plan quality and coverage remain inconsistent. In a 2018 study, Jacob W. Malcom and Ya-Wei Li concluded that ESA recovery plans were "missing, delayed, and old." They found that one-quarter (24.5 percent) of listed species still lacked recovery plans and that of the plans that had been completed, half had taken until more than five years after listing to finalize, half were more than twenty years old, and the treatment of different taxonomic groups was inconsistent. In total, 379 out of 1,548 species that had been listed for at least 2.5 years and were eligible for recovery plans in fact lacked recovery plans.[36] Today, 1,378 of the 1,668 listed U.S. species have active recovery plans.[37] A further study published by Katherine Wright and Shawn Regan in 2023 highlighted the variability of recovery plan *timelines*. They found that among those few species for which the FWS ventured to estimate when recovery would be complete, only 3 percent of their estimates were accurate.[38] Clearly, recovery planning is an area in need of improvement.

The American peregrine falcon (*Falco peregrinus anatum*) is one of three sub-species of peregrine falcon native to the United States. Listed as endangered in 1970, the falcon was delisted in 1999, following the 1972 banning of DDT and a decades-long captive breeding and reintroduction program led by private organizations. *Frank Doyle/U.S. Fish and Wildlife Service National Digital Library*

When the peregrine falcon was listed in 1970, it had been reduced to forty breeding pairs in the western United States and extirpated in the East. By 1999, the species had recovered to 1,425 breeding pairs nationwide.[39] Since delisting in 1999, the status of the peregrine falcon in the United States has continued to improve. By 2003, the end of the statutory five-year post-delisting monitoring period, the FWS identified 3,005 breeding pairs in North America, a 70 percent increase since delisting.[40]

To wildlife advocates, the value of the ESA-mandated federal monitoring period is clear. There are numerous species, ranging from the gray wolf in the northern Rocky Mountains to the peregrine falcon, for which population figures through the statutory five-year monitoring period are readily available from the FWS but which become much harder for the public to keep track of beyond five years after delisting.

The 1988 amendment to the ESA required that the Services monitor species for five years after delisting—"in cooperation with the States."[41] But in practice, it sometimes became an unfunded mandate on states and an impediment to delisting species, increasing the friction between state and federal wildlife managers. In notable cases such as the gray wolf in Wyoming, scrutiny of the state's management plans for the wolf led to extensive controversy, delays in delisting, and litigation. In purely philosophical terms, some states take offense at the continued federal imposition on their sovereign authority over wildlife, which they believe should be fully returned to them when a species is delisted.

CHAPTER 6

The Apex Predator Problem

WOLVES CAPTURE THE IMAGINATION AND HEARTS OF MANY, BUT FOR others, this apex predator is a stressful daily threat to their domestic and farm animals. For more than a century, westerners wary of wolves' predatory strikes hunted, trapped, poisoned, and even incinerated them out of existence. In their defense, they were supported by federal policy, federal wolf extermination programs, and federal wolf bounties. But during the Green Revolution, one of the first species to be protected by the ESA was the gray wolf, which was listed in 1967 under the earlier 1966 act.

Nearly three decades later, in 1995, the federal government relocated fourteen wolves from Canada to Yellowstone National Park and fifteen to central Idaho. In 1996, seventeen more were released in Yellowstone and twenty more in central Idaho.[1] It took exactly one week for the first domestic livestock calf to be killed by a wolf released in Idaho, which was in turn shot dead, and the population has been embroiled in controversy ever since.[2]

Local residents by and large do not love wolves. According to the National Park Service, the largest number of public comments ever received on any federal proposal at that time—more than 160,000—were received and recorded. Between 1995 and 2003, 256 sheep and 41 cattle were killed by wolves outside Yellowstone, as locals were asked to be patient and adapt for the benefit of the ecosystem.[3]

By 2008, the FWS under President George W. Bush made a scientific determination that the population had recovered and delisted the gray wolf in Montana, Idaho, and Wyoming, returning the species to state management, including the regulation of hunting. Immediately, twelve conservation groups sued and won a preliminary injunction against the FWS's delisting. After a few months, the court allowed the delisting to proceed, though Bush's FWS did not delist the wolves in Wyoming, opining that state's management plan wasn't as comprehensive as the other states with wolf populations.

But that wasn't sufficient for the litigious groups. In 2009, with the FWS under the control of the new Obama administration, they sued again—and won. In 2010, a U.S. district court judge ruled that the FWS's delisting rule did not comply with the ESA in Idaho and the northern Rocky Mountains. The wolves that had been delisted were returned to the endangered species list.

Sixty-six gray wolves captured in Canada were released in Yellowstone National Park and central Idaho in 1995–1996. In this January 1995 photo, the first Canadian wolf is introduced to the Crystal Bench wolf acclimation pen in Yellowstone National Park's Lamar Valley. From left to right: Mike Phillips, Jim Evanoff, U.S. Fish and Wildlife Service Director Mollie Beattie, Yellowstone Superintendent Mike Finley, and Secretary of the Interior Bruce Babbitt. *Jim Peaco/National Park Service Photo Gallery*

Clearly, the ESA's 1982 amendment expanding citizen enforcement of the ESA had created an endless tug-of-war between the executive branch and the judicial branch. The executive branch, regardless of whether its philosophy was conservative like Bush's or liberal like Obama's, could not loosen restrictions without bringing a lawsuit on itself. Lawsuits against the Bush administration were somewhat understandable. An inspector general's report found that Bush's deputy assistant secretary, Julie MacDonald—who was deeply sympathetic to the oil industry and property rights groups—had exerted undue political influence and overridden FWS scientists, jeopardizing fifteen endangered species listing decisions.[4] And Bush's deputy secretary of the interior, Steven Griles, was convicted and spent ten months in jail for lying to the Senate Indian Affairs Committee.[5]

But the litigious conservation groups showed no signs of letting up under the much greener Obama administration. The legislative branch would need to intervene. In 2011, Congress delisted wolves in Montana and Idaho in a rider on an unrelated piece of legislation, and the FWS under Obama proposed delisting in Wyoming.

The coalition gave notice: they would sue yet again. "It's extremely disheartening to watch the Obama administration unravel one of our country's greatest Endangered Species Act success stories by turning over the conservation of wolves to states such as Wyoming and Idaho that treat these animals like unwanted vermin," said Mike Senatore from the Defenders of Wildlife organization, one of the organizations in the coalition along with the CBD.[6]

Tongue in cheek, the Wyoming state legislature considered a resolution taking aim at coastal elites like these, who take it on themselves to make decisions for people whose daily lives and livelihoods are impacted by wolves and other predators: "WHEREAS, in the event Congress authorizes the creation of the Central Park Wilderness, the state of Wyoming shall donate a breeding pair of grey wolves to begin the recolonization of the grey wolf to Manhattan, an area, like Wyoming, encompassing the grey wolf's historic range."[7]

A federal district judge ruled in favor of the environmental coalition and against Wyoming in 2014, opining that the state's commitment to maintain its wolf population lacked safeguards and that Obama's FWS shouldn't have accepted it. Wyoming's wolves were back on the endangered species list until 2017, when a federal appeals court reversed the lower court's ruling. The gray wolf has been delisted in Wyoming ever since.

Although rightly regarded as lighthearted by all involved, Wyoming's 2012 resolution correctly pointed out that the island of Manhattan—the heart of New York City and home to 1.7 million people living 75,000 to the square mile—is within the gray wolf's historic range (as is virtually the entire United States) and likely had a native population of gray wolves in the distant past. Using that standard alone, Central Park is as suitable as Wyoming for a wolf population. Most Manhattanites wouldn't tolerate wolves in such close proximity; why should Wyoming residents have to? How far are we willing to go to restore a species or ecosystem?

No ecosystem in the United States is untouched by the impact of humans. We can't restore ecosystems to some idyllic prehuman past no matter how hard we try. For thousands of years before Europeans arrived in the Americas, we are learning that the native peoples manipulated landscapes by setting fires. So we can't even imagine what our prehuman ecosystems might have looked like. All we can do now is conserve the biodiversity that remains and restore it for the benefit of humans as well as wildlife. But what does that restoration look like?

Some places are too densely populated for wolves. But, for example, they may be suitable for backyard gardens to support native pollinators. How do we decide where to draw these lines? Perhaps just as important, *who* gets to decide where to draw these lines? The ESA calls for these decisions to be made by the federal government through regulations and in consultation with state governments and with public input and at all times for the best available science to guide these decisions.

All the same, there are flexibilities, nuances, and policy decisions that can be made only by people and that not all people will necessarily agree with. The ESA is not some abstract, philosophical, aspirational legislation. It is a practical law with very real implications and impacts, both positive and negative, that America must grapple with anew with each passing year. No situation better encapsulates this immutable truth than the choices, changes, and impacts that inevitably—if sometimes unpredictably—ensue when apex predators put their paws on the ground and—with human assistance—reclaim some small part of their ancient domains.

THE COST OF PREDATORS

The Mexican gray wolf, or lobo (Spanish for "wolf"), *Canis lupus baileyi*, is the smallest and most imperiled subspecies of gray wolf. Extirpated from the United States by the time of its separate listing in 1976, breeding of wolves captured in Mexico allowed a small population to be reintroduced more than twenty years later in 1998, and today 242 wolves (as of the end of 2022) can be found roaming rural Arizona and New Mexico. They occupy a designated "recovery zone" bounded on the south by Interstate 10 and on the north by Interstate 40, some 200 miles away. Wolves that wander beyond those borders are subject to capture and relocation back into the primary wolf release area, which spans Apache National Forest in Arizona and part of New Mexico and the Gila National Forest in New Mexico, a total of some 4.4 million acres.[8]

These wolves are the subject of intense controversy, as Wyoming's were. The most colorful, widely reported incident occurred in Reserve, New Mexico (population 289), nestled in the Apache National Forest, nearly a decade after the reintroduction began. The town installed some twenty wooden and mesh cages, ostensibly to protect schoolchildren from ravenous wolves that might attack them as they waited for their school buses. The cages were ridiculed by wolf advocates and the national press, who regarded the notion that skittish wolves would attack schoolchildren in broad daylight as absurd. Indeed, in the more than twenty years since the cages were built, there have been no documented cases of Mexican wolves attacking people (although on rare occasions around the world, wolves have attacked humans—especially when afflicted with rabies).[9]

Cattlemen and sheep growers view wolves of all sorts—and other large predators—as threats to their livelihoods, and for good reason. Wolf depredations of livestock are common and widely documented. When they are reported, state and federal wildlife authorities investigate and verify the loss. Offending animals may be relocated or killed. Payments are made to livestock operators to make up for their losses and to encourage them to tolerate wolves.

The endangered Mexican gray wolf (*Canis lupus baileyi*) was first reintroduced to Arizona and New Mexico in 1998. State and local leaders issued dire warnings that wolves would prey on children playing outside and later protested the expansion of the species' recovery area. In 2015, the state of New Mexico denied the U.S. Fish and Wildlife Service a permit to continue releasing wolves, which the federal agency ignored, resulting in litigation and further acrimony. Most recently, environmentalists have successfully used public pressure and litigation to push the federal government into increasing population targets for the species. These state–federal and perceived human–wolf conflicts have hindered recovery of the species, and its future remains in doubt. *Jim Clark/U.S. Fish and Wildlife Service*

The scale of these programs is considerable. For example, in the northern Rocky Mountains around Yellowstone National Park, for twenty-three years, the environmental group Defenders of Wildlife operated a program that paid livestock growers for wolf depredations, covering verified wolf-caused losses of livestock at 100 percent of fair market value (up to $3,000) and probable losses at 50 percent of fair market value (up to $1,500). The program wound down in 2010, replaced by a new federal program authorized by the Omnibus Public Lands Management Act of 2009, by which time it had paid livestock operators more than $1.4 million.[10] Under the continuing federal program, approximately $500,000 per year is spent on depredation compensation grants and an equal amount on depredation prevention grants.[11] States within wolf (or grizzly bear) range offer livestock depredation payments as well.

In 2007, the town of Reserve, New Mexico, began building cages at rural school bus stops in a publicity stunt supposedly intended to protect schoolchildren from attack by Mexican gray wolves. The cages attracted national media attention and effectively showcased the rural community's deeply felt objections to wolf reintroduction. *Photograph by Laura Paskus*

Widely cited data show that wolves kill only 1 in 44,853 cows, fewer than are killed by domestic dogs and other native carnivores. Moreover, all deaths from carnivores (including birds) are a tiny fraction of total deaths, far outweighed by mortality from ill health, weather, birthing problems, poisoning, theft, and so on.[12] The numbers are worse for sheep, which are killed by wolves at a rate of 1 in 7,193, but the overall losses are too small to be economically damaging to the industry.[13] But these statistics and the environmental groups and wolf advocates who publicize them downplay the severity of the issue.

According to ranchers (and a growing body of research), actual depredations are only the tip of the iceberg. In addition to direct predation, wolves can cause stress in domestic livestock, leading to lower birth rates and slower weight gain, both of which impact livestock operators' bottom lines.[14] Raising livestock can be a very marginal business, especially for small operators, and wolf impacts can be distributed unevenly, so wolf depredations can have a meaningful negative impact.[15] As a result, the issue is taken seriously by all involved, with significant state, federal, and private efforts to reduce wolf depredations and compensate livestock operators for them in order to manage and reduce human–wolf conflicts.

LOBO LITIGATION

The impact of wolves on livestock operators, the locations of wolves in mostly conservative states, and the traditional alignment of the livestock industry with the Republican Party have all given rise to extreme partisanship around wolves and wolf management. Opposing wolves and decrying their impact on people (as well as fearmongering their potential danger to children) has become a cause célèbre on the political right. And celebrating wolves, wanting wolves in more areas, and filing lawsuits to keep wolves on the endangered species list at all costs is equally popular on the left. This pattern repeats across the country.

In the case of the Mexican gray wolf, the initial reintroduction attracted relatively little attention despite the anxiety in the town of Reserve. Nearly two decades later, in 2015, the Obama administration greatly enlarged the primary Mexican gray wolf recovery area and increased population targets from at least 100 wolves to 300 to 325.[16] This new, stronger rule was challenged by environmentalists who felt it did not go far *enough*. They secured a court ruling against the rule, holding that it "only provides for the survival of the species in the short term and therefore does not further recovery."[17]

In response, the Biden administration changed the population goal from a target of *between* 300 and 325 to a *minimum* of 320 wolves, established genetic diversity targets, and restricted allowable take of Mexican gray wolves until the genetic diversity targets were met.[18] This new plan remains the subject of an environmentalist lawsuit, so just how far the government will be forced to go in furthering Mexican gray wolf recovery remains to be seen.[19]

WOLVES AND BIODIVERSITY

If large predators cause damage to livestock businesses and inspire fear and controversy among people, what are their benefits? The answer is that they are essential components of healthy, functioning ecosystems and that they help sustain biodiversity. Specifically, apex predators regulate ecosystems in a process known as a "trophic cascade," reducing the numbers of their prey species and ensuring that they do not become overpopulated. All parts of the web of life move in relation to each other: in a wet year, more plants allow there to be more herbivores and, in turn, more predators. When predators consume too many prey species, their food supply is diminished, and their own numbers decline, allowing prey populations to recover. Each species moves in cycles related to weather and other outside factors and in relation to each other.

A classic example of the devastation that can be caused by the *lack* of predators arose in the Kaibab Plateau of Arizona in the 1920s. The legendary ecologist Aldo Leopold—then a young forester early in his career with the U.S. Forest Service—surmised that a campaign of predator control of the coyote, mountain lion, wildcat, and wolf and bear populations would benefit mule deer. In fact, the opposite happened. Freed of predation, the mule deer population exploded, exceeding the carrying capacity of the habitat, and needed to be reduced through human intervention. At the time, the complexities of the habitat's carrying capacity, deer population, and reproduction dynamics were not fully understood.

A modern example of the benefits of apex predators has unfolded in the northern Rocky Mountains in and around Yellowstone National Park since the reintroduction of gray wolves to the area in 1995. Predation of elk by wolves has resulted in an increase in the biodiversity of the ecosystem. Yellowstone now has more aspens, healthier willow stands, increased beaver colonies, decreased erosion, and other associated ecological benefits that have not been studied and quantified.[20] This result has become part of the accepted narrative of wolves in Yellowstone and is widely cited by environmental advocates as a reason why wolves should continue to be protected.[21]

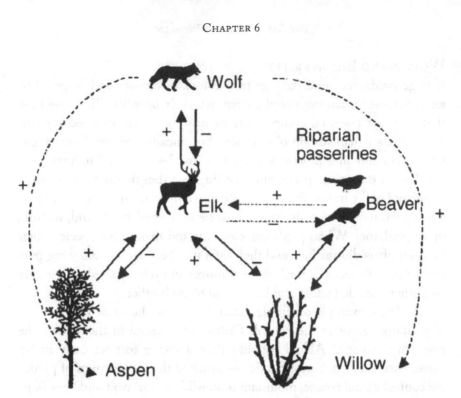

This illustration shows simplified trophic interactions in the Bow Valley of Banff National Park, Alberta, Canada, which hosts many of the same species as Yellowstone National Park in the United States. Solid lines represent direct interactions, while dashed lines indicate indirect effects. + indicates an increase in a population, while − represents a decrease. *From Mark Hebblewhite, Clifford A. White, Clifford G. Nietvelt, John A. McKenzie, Tomas E. Hurd, John M. Fryxell, Suzanne E. Bayley, and Paul C. Paquet, "Human Activity Mediates a Trophic Cascade Caused by Wolves,"* Ecology 86, no. 8 (August 2005): 2138.

Of course, the purpose of the reintroduction of wolves to Yellowstone was to recover the species under the ESA, not to regulate Yellowstone's ecosystem.[22] And the subject of fierce debate in the scientific community today is whether wolves have, in fact, further established "an ecology of fear" in Yellowstone. In an ecology of fear, the mere *presence* of predators affects the *behavior* of prey species.[23] In the case of Yellowstone, an elk that is afraid of wolves might move more often, spending less time in its favorite browsing areas, such as riparian valleys, and more in other habitats, reducing browsing pressure within its preferred habitats.

The theory that wolf reintroduction established an ecology of fear in Yellowstone was once widely accepted, but recent studies have found that elk do *not* modify their behavior in the presence of wolves. One possible explanation for this lack of response is that the wide-ranging and elusive wolves are effectively everywhere, and so elk respond by ignoring them.[24] In the case of Yellowstone, this is largely an academic question. Direct predation of elk alone has been enough to cause a trophic cascade and the associated ecosystem benefits.

But it has potential importance elsewhere, such as for the Mexican gray wolf. There are only 242 Mexican gray wolves in the wild, in contrast to 2,440 gray wolves in the northern Rocky Mountains. If wolves cannot establish an ecology of fear, there may be too few Mexican gray wolves to have much impact on the ecosystem in Arizona and New Mexico.

THE WOLF WATCHING INDUSTRY

Yellowstone National Park visitors love wolves, and wolf watching is a major industry. Whenever a wolf pack is sighted, word spreads, and wolf watchers—distinctive for the powerful spotting scopes they set up—quickly congregate. The National Park Service estimates that wolf watchers traveling to Yellowstone contribute $35 million per year in economic activity in the region.[25]

Notwithstanding the ever-present controversy, the northern Rockies has been a resounding success for the wolf. Not only have wolves improved biodiversity and spurred tourism, but they have also flourished, increasing from 66 to 2,440. The population reached its recovery target of thirty breeding pairs and 300 wolves for three consecutive years in 2002 and should have been delisted at that time. Their delisting was prevented by environmentalist lawsuits that were decided against the FWS in 2005, 2008, and 2010.[26]

Since delisting, a new controversy has arisen around the northern Rockies: hunting. Wolves have been subject to hunting in some areas since 2009, with no ill effects on the overall population.[27] But in 2021, the legislatures of both Idaho and Montana passed laws requiring their respective fish and wildlife agencies to increase the number and means of hunting wolves in order to reduce their overall wolf populations. This decision led to an outcry from the environmental community,[28] and two groups filed petitions requesting that the gray wolf be again listed under the ESA on an emergency basis.[29] Although the FWS did not list the species on an emergency basis, it announced on September 17, 2021, that it would initiate a twelve-month review of the status of the species.[30] As of September 2023, two years later, that review is ongoing. Depending on the outcome of that review, the wolves in the northern Rockies population may again be listed under the ESA.

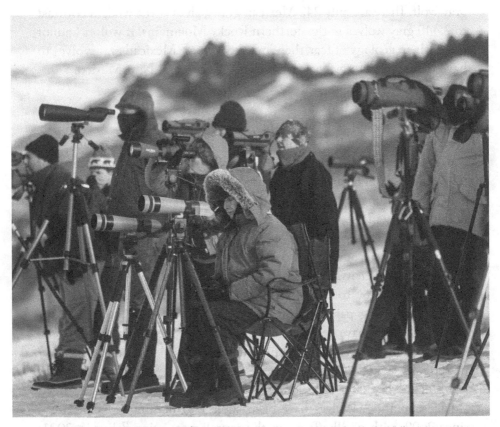

Wolf watching has become a $35-million-per-year tourist industry in Yellowstone National Park, supporting outfitters who rent powerful spotting scopes, cameras, and other equipment as well as hotels, restaurants, and other businesses in the area. *Sean Sperry/Bozeman Daily Chronicle*

THE WESTERN GREAT LAKES: ENDLESS LITIGATION

A similar story has played out in the western Great Lakes states, home to the only surviving native (as opposed to introduced) population of wolves in the lower forty-eight states. Because of the large number of wolves in Minnesota, in 1978 that population was reclassified from endangered to threatened and is now managed as the Western Great Lakes Distinct Population Segment. Protected by the ESA, this native population thrived: by 2001, it had reached its target population of 1,600 and was ready for delisting, and the 2020 population reached 4,222.[31] But like the northern Rockies population, litigation by environmental groups stymied its delisting. Judicial decisions against the FWS's delisting rules were issued in 2005 (twice), 2008, 2009, and 2014.[32]

The 2014 western Great Lakes case led to an appeal and a 2017 ruling with far-reaching consequences. On appeal, the Court of Appeals for the D.C. Circuit affirmed the ruling against the government but limited its ruling to a narrow rationale: that the delisting rule failed to properly account for the future status of the "remnant" population of wolves outside the "distinct population segment," which would become unprotected.[33] The remnant issue arose due to a peculiarity of distinct population segments. In earlier cases, courts had resisted the notion of ever delisting one on its own because to do so could enable a sinister workaround that would undermine the ESA. The ESA provides for distinct population segments in order to *protect* isolated imperiled populations of otherwise healthy species. But in the case of a species like the wolf, where there is an isolated *healthy* population of an otherwise extirpated species, one could imagine a scenario in which the government might draw lines around small populations and declare them recovered, declare the species elsewhere extinct, and delist the entire species without ever actually grappling with the possibility of recovering it. And recovering it is what the ESA requires.[34]

In response, the FWS adopted a new approach, proposing to delist the species nationwide. By delisting the species nationwide, the agency sought to avoid the issue of a remnant population outside of a delisted distinct population segment. The rule accomplishing this was finalized on November 3, 2020, with an effective date of January 4, 2021.[35] A little over a year later, on February 10, 2022, a federal district court struck down the nationwide delisting rule, once again listing the western Great Lakes population as a threatened species, which it remains today.[36]

During the period the wolf was delisted in 2021, this population also became the subject of a fierce controversy over hunting. In February 2021, Wisconsin held a wolf hunt in accordance with a state law requiring an annual wolf season once the species was delisted. Ordinarily, wolf hunting in Wisconsin would occur from November through February, but after the species was delisted in January, hunting organizations won a state court lawsuit that required an abbreviated—and rushed—hunt in February.[37] The state's Department of Natural Resources duly complied and set a target of 119 wolves, but, in just three days, hunters instead killed 216 wolves, almost double the quota.[38] The intensity of wolf taking by hunters during the denning season created uncertainty about the overall wolf population in the state, and biologists with the Department of Natural Resources recommended a limit of 130 wolves for the 2021–2022 hunting season. The state's Natural Resources Board disregarded this recommendation and voted to set a quota of 300 wolves, and both that decision and the state's wolf-hunting law became the subject of litigation.[39] On October 22, 2021, a state judge declared that the department had unconstitutionally misapplied the state's wolf hunting law and enjoined the 2021–2022 wolf hunt.[40] Thereafter, the western Great Lakes population was again listed under the ESA, rendering the issue moot—at least for the time being.

GRIZZLY BEARS

The saga of the listing status of the gray wolf and the accompanying litigation has been echoed in an eerily similar controversy over the *other* large mammalian predator native to the lower forty-eight states that is protected by the ESA: the grizzly bear—particularly the Greater Yellowstone Ecosystem's Distinct Population Segment, which is also ready to be delisted. This population has played a central role in the history of Yellowstone National Park. In fact, in the early twentieth century, the ability to watch bears feed on leftover food was a major factor in the park's growing popularity among tourists.[41] By 1975, however, when the grizzly

Human feeding of bears in Yellowstone National Park was a major tourist attraction for decades, helping to build the public's interest in the park and also helping sustain a dense population of bears before it was phased out in the late 1960s.
Yellowstone's Photo Collection/Yellowstone National Park, National Park Service

bear was listed as a threatened species under the ESA, the population in and around the park had dropped to 136 bears. This was the southernmost and most isolated of five populations in the lower forty-eight states, which together had no more than 800 to 1,000 bears among them.[42]

Working in cooperation with state fish and wildlife agencies through the Interagency Grizzly Bear Committee (established in 1983), hunting seasons for bears were closed, bear habitat was protected from disruptions such as roadbuilding and domestic livestock grazing, public education campaigns sought to reduce human–bear conflict, and funds were established to compensate domestic livestock operators for animals lost to bears. The bear population responded, and by 2007 the Greater Yellowstone population had reached all three of its recovery goals: there were at least 500 bears, there were female bears with cubs born that year in sixteen of eighteen designated bear management units, and there was a stable or increasing population.[43] The population's growth has continued since: as of 2022, estimates indicated that 965 grizzlies lived in the Greater Yellowstone Ecosystem Distinct Population Segment.[44]

The grizzly bear (*Ursus arctos horribilis*), listed as a threatened species since 1975, has been recovering under the management of the Interagency Grizzly Bear Committee, established in 1983. The Greater Yellowstone Ecosystem distinct population segment has been ready for delisting since 2007, but lawsuits have intervened. In addition to the Greater Yellowstone Ecosystem, there are four other extant bear populations: the Northern Continental Divide Ecosystem, which has also reached its recovery targets; the North Cascades Ecosystem; the Selkirk Ecosystem; and the Cabinet-Yaak Ecosystem. A sixth bear recovery zone, the Selway-Bitterroot Ecosystem, is currently unoccupied. *Jean Beaufort/www .publicdomainpictures.net*

When the Greater Yellowstone Ecosystem Distinct Population Segment first reached its recovery goals in 2007, a delisting rule was issued.[45] In 2011, however, a federal district court invalidated the delisting because FWS had not adequately considered the potential impact of declines in whitebark pine, a food source for the bear population, which was suffering from climate change, wildfires, and disease.[46] On appeal, the Ninth Circuit upheld the ruling.[47] The FWS and the Interagency Grizzly Bear Committee then undertook an extensive study of whitebark pine, ultimately concluding that potential loss of whitebark pine was not a limiting factor for grizzly bear recovery.[48]

Thereafter, the FWS again issued a delisting rule for the Greater Yellowstone population of grizzly bears on June 30, 2017.[49] However, a month later, on August 1, 2017, the D.C. Circuit issued its "remnant" ruling preventing the delisting of the western Great Lakes population of gray wolves.[50] The grizzly bear delisting rule suffered from the same apparent defect. After review, the FWS determined that there would be no issues with unprotected remnant populations of the bear,[51] but on September 24, 2018, a federal district court in Montana disagreed.

It further held that the FWS violated the ESA's "best available science" standard by agreeing—as a concession to state governments—that in the future, the Greater Yellowstone population could be managed according to existing population models rather than recalibrating population models in the future as new science emerged.[52] On appeal, the Ninth Circuit disagreed with the district court regarding the remnant population but affirmed its decision on the basis of the recalibration issue.[53] As a result, as of 2023, the Greater Yellowstone population of grizzly bear remains listed.

THE BLACKFOOT CHALLENGE: MANAGING PREDATOR CONFLICTS THROUGH PARTNERSHIP

In the Blackfoot River Valley of Montana, made famous by the 1976 novella and 1992 film *A River Runs Through It*, a new model has emerged for state and federal government agencies and landowners to work together to reduce and manage conflicts between both grizzly bears and gray wolves and their livestock. The Blackfoot Challenge, a landowner-led collaborative conservation organization with roots in the 1970s, formally incorporated in 1993, has been the key. The organization's mission is to "coordinate efforts to conserve and enhance natural resources and the rural way of life in the Blackfoot watershed for present and future generations."[54]

It provides "a forum to encourage civil dialogue to support environmentally sustainable stewardship of the [Blackfoot River] watershed through cooperation of private and public interests."[55] The organization believes that "effective partnerships and working relationships are based on trust, respect, credibility, and the ability to empathize across a diversity of values," and it takes a strict non-advocacy and non-litigation approach to addressing local problems through which it has, over time, earned the trust and support of many local residents.[56] Stretching back to the 1970s, Blackfoot Valley landowners have supported numerous conservation programs in the watershed, including stream and wetland restoration, native fish restorations, access for hunters and anglers and other public recreation, noxious weed mitigation, timberland restoration, and more than 100,000 acres of conservation easements.[57] With this history of successfully managing public–private projects around issues such as forest management, irrigation improvements, and wildlife reintroductions, the Blackfoot Challenge was an obvious leader for the community to turn to when grizzly bears from the Northern Continental Divide Ecosystem began arriving in the area in the late 1990s.[58]

When a hunter was killed by a grizzly bear in 2001, public attention focused on the (heretofore nonfatal) human–grizzly conflicts that had been occurring in the Blackfoot Valley since 1998. The Montana Department of Fish, Wildlife, and Parks then approached the Blackfoot Challenge for assistance. In response, the group formed a Wildlife Committee, eventually including forty-five members, including state and federal agencies, environmental organizations, and research universities. The group began its work by surveying local ranchers and outfitters—listening to the local community—to understand their perspectives on grizzly bears and how they might coexist with bears. This approach enabled the group to set priorities for its work that reflected the needs and values of the local community, which included protecting human safety, protecting private property from bear damage, and protecting rural livelihoods.[59]

The Wildlife Committee worked toward these goals through an iterative process of science, communication, and participatory projects in which it surveyed the local community and held public meetings, then mapped and analyzed potential sources of conflict on the landscape, communicated its findings back to the local community, and adjusted its planning based on their concerns. Through the process, a number of programs emerged, including electric fencing to protect newborn calves from bear depredation, collection and composting of livestock carcasses to reduce bear attractants, more general sanitation and bear-proofing, outreach and education programs, and a neighbor-to-neighbor communication system.[60] Each of these required community support and involvement and were designed to address community concerns. In the case of fencing, the Wildlife Committee coordinated project implementation; federal agencies, state agencies, the Blackfoot Challenge, and private foundations provided funding; and ranchers provided in-kind donations through labor to remove traditional fences. (More recently, electric mats have been added to the predator exclusion toolbox.[61]) In the case of carcass removal, the pickup service is provided by the Blackfoot Challenge, and the Montana Department of Transportation provides composting of the carcasses. Because western ranchers are private people, averse to disclosing their financial positions—including livestock mortality figures—the program offers anonymous drop-off points for carcass disposal.[62] Participation in the program has increased over time, and it now collects between 450 and 550 livestock carcasses per year.[63]

The results of the Blackfoot Challenge's grizzly bear program have been striking. There was a 74 percent reduction in verified human–bear conflicts within their project area between 2003 and 2013 and a downward trend in known grizzly bear mortalities—even as the bear population increased by approximately 3 percent per year.[64] And between 2013 and 2019, there were just 2.6 average annual confirmed and probable livestock losses to grizzly bears out of a population of 16,000 to 18,000 head of livestock.[65] Thanks to the Blackfoot Challenge's engagement with landowners, wildlife managers received more accurate information on when and where bears appeared, generating data that they could use to improve management.[66] In addition, there was an increase in trust and credibility among the many stakeholders, economic benefits for livestock operators whose losses were reduced, and an increase in community acceptance that grizzly bears had come to their valley to stay.[67]

The Blackfoot Challenge's success with grizzly bear management built on its past record in the community and set the stage for a similar program to address gray wolves, which moved into the area in 2007 and within five years increased their numbers from one local pack to twelve.[68] Established programs such as livestock carcass removals and electric fencing helped address wolf management issues. In addition, the group conducted a range survey with the assistance of local ranchers and then implemented additional management tools, such as fladry (bright ribbons on wires to dissuade wolves) and range riding (the practice of increasing livestock herd supervision rates and monitoring of wolves), in order to protect livestock, better understand wolf activity, and regularly communicate with livestock producers about the status of their livestock in relation to wolf activity.[69] Between 2006 and 2015, confirmed livestock losses to wolves averaged 2.2 depredations per year, and an average of 2.4 wolves were removed each year in response.[70] Both these figures were lower than in other areas of Montana, suggesting that the Blackfoot Challenge's proactive wolf management activities were helpful—although the organization's scientists stress that they do not have proof of causation.[71]

The gray wolf (*Canis lupis*) historically ranged across North America. First listed in 1967, the gray wolf has generated more controversy and litigation—and stronger emotions—than any other endangered species. *Gary Kramer/U.S. Fish and Wildlife Service*

The Blackfoot Valley's experience with grizzly bears and gray wolves is an example of successful collaborative conservation and a model for other groups and regions where large carnivores are present or may be reintroduced. A landowner-led group turned the collective abilities of its members and partners toward solving a common problem: the coexistence of large predators with commercial livestock operations. It took the support and direct involvement of local landowners that helped develop programs that served their needs while also relying on scientific research techniques to fit those programs to the landscape and the needs of the species. Finally, it engaged partners in federal government, state government, the NGO community, and universities to broaden its impact and leverage more resources and expertise. All of this was accomplished through trust, collaboration, and sound science, with no room for partisanship, mythology, or fearmongering.

PREDATORS AND THE ESA

The gray wolf has a population of 7,000 to 11,000 in Alaska (where it has never been protected as a threatened or endangered species) and 200,000 to 250,000 worldwide. The IUCN regards it as a species of least concern.[72] Another species of least concern is the grizzly/brown bear, 58,000 of whom live in one interconnected, continental population that stretches from the northern Rocky Mountain states through Canada to Alaska. (The IUCN lists grizzly and brown bear subspecies and populations individually, and it considers the Yellowstone grizzly bear population "vulnerable."[73]) Clearly, wolves and bears are not being listed and conserved—at great expense and controversy—to preserve a unique and irreplaceable genetic heritage that might someday provide humans with an important source of medicine or research. The world has abundant resources of both wolves and bears for that purpose.

The protection of wide-ranging and controversial apex predators such as wolves and grizzly bears invites important questions about the ESA: Why are we protecting these species, courting political controversy, and investing millions of dollars to establish new populations? What lengths should we go to in the name of conservation?

The simplest answer is that we are conserving these species (and establishing new populations of them) in the lower forty-eight states because the ESA commands us to do so for the sake of Earth's biodiversity. Its purpose is to conserve species and "the ecosystems upon which [they] depend." The conservation of apex predators enhances biodiversity and, in turn, ecosystems.

On its face, and according to many environmental advocates, the ESA demands that we go as far as humanly possible. But in practice, the ESA is flexible and its application practical. We have seen this in many other examples in this book, and in the case of the wolf or the bear, we see it again in the relatively small number of places where we have chosen to reintroduce these species.

In 2020, voters in Colorado adopted Proposition 114, requiring the state to reintroduce and manage a resident population of gray wolves. At that time, wolves from the northern Rockies population were only just beginning to reach Colorado. Although lone wolves had traveled in and out of the state over the years, the first resident wolf was identified in 2019, the first pack in 2020, and the first breeding pair in 2021.[74] Proposition 114 was controversial and bitterly contested, passing by a margin of 50.91 to 49.09 percent in a Colorado general election vote reflecting the state's growing urban–suburban divide: support for wolf reintroduction came largely from the urbanizing counties along the Rocky Mountain Front.[75] The debate over the proposal played out along familiar lines: supporters praised the potential for wolves to improve Colorado's ecosystems as an apex predator, while opponents decried wolves' impacts on Colorado's livestock and big-game industries.[76]

Colorado's Wolf Restoration and Management Plan was finalized on May 3, 2023. As wolves are an ESA-listed species, Colorado Parks and Wildlife is now working with the FWS to develop a rule to allow the establishment of an experimental population of gray wolves in Colorado. They will then translocate approximately thirty to fifty wolves over a period of three to five years, most likely from the nearby northern Rockies population.[77] The wolf reintroduction effort is already the subject of legislative attacks and lawsuits both for and against it and is sure to remain controversial for years to come.

Like the other mountainous states to its north and south, Colorado has ample suitable habitat for wolves. Wolf advocates dream of someday establishing one interconnected population that spans the Rocky Mountain states. But their ambitions have limits. Landscape-scale restoration of species and ecosystems is expensive and brings with it both controversy and opportunity costs.

CHAPTER 7

The Price Tag for the ESA

"YOU JUST DO NOT SAVE THE SPECIES WITH GOOD WISHES . . . I CAN SLUG somebody with the cost of buying habitat or paying for protection,"[1] Subcommittee on Fisheries and Wildlife Conservation and the Environment chairman John Dingell warned Sierra Club lobbyist Robert C. Hughes in a 1973 hearing on his ESA legislation.

Indeed, even if wolves and bears weren't a threat to livestock, and even if every other conflict arising from the ESA were to disappear, stubborn facts would remain. For one, as many as 84 percent of ESA-listed species require ongoing human manipulation of their habitats to sustain their populations.[2] For these conservation-reliant species, it is not enough to protect them from hunting or to establish a preserve where their habitat will be protected. Instead, their habitat must be managed by humans— for example, by our removal of the predators who kill them or by our burning forests so that they have new-growth trees essential to their survival. For these species, the quality, durability, and fiscal soundness of long-term conservation plans is critical. Our investments in conservation must be sustained over time lest they be wasted.

Dingell had no patience for people with big ideas about saving the environment but no way of footing the bill. And while groups like the Sierra Club might have been loath to admit it, until the Green Revolution, sportsmen were largely footing the costs of conservation. That's mostly true today as well. "As you know, the hunter buys the duck stamps, the fisherman buys a fishing license, but the dickybird watcher raises cain about the hunters slaughtering the beast. But they have not put a nickel into the pot, and this is a source of constant irritation to me, and the game management people, who have been paying the freight for these many years, while these other people have been out fussing," Dingell boomed in frustration during his bill's hearing.[3]

During the debate leading up to the ESA's passage, states called for a combination of state and federal resources deployed cooperatively. The testimony of many compared the states' capacity to the Bureau of Sport Fisheries and Wildlife (BSFW) within the FWS, which would administer and enforce the ESA. States had 5,800 full-time law enforcement officers, 32.3 percent of the states' combined employees, and 27 percent of state agencies' budgets; $72 million was spent on enforcement alone by the states in 1972 versus $5.5 million by the BSFW. The federal bureau had 3,000 employees in total, including for law enforcement, science, and clerical work, among other functions. The states had more than 5,400 biologists alone and 18,000 total employees. States' manpower and financial resources could administer a national endangered species program—but only provided that they received federal funding to supplement state funding for the new species recovery programs that would be required.

Ten states all provided testimony or letters that they wanted "concurrence" in both listing and enforcement decisions, not just "consultation" at the secretary of the interior's discretion. They simply wanted no infringement of their historical and traditional states' rights over managing resident wildlife. However, management of migratory and international wildlife, they conceded, was a federal responsibility.[4] They were prepared to have specific criteria established by their democratically elected members of Congress (not appointed officials in the Department of the Interior) for the uniform protection of endangered—not merely threatened—species against which state programs could be measured. The federal role would be one of oversight, looking over the states' shoulders. If a state could not perform, the federal government would assume authority that states would acknowledge.[5]

The states planned to build off the programs then in place based on the proven scientific principles of wildlife management. The Nixon administration was open to this approach, proposing that the federal government "may include consultation" with the states at the secretary's discretion.

Congressman John D. Dingell Jr. (D-MI) followed his father, who served twenty-two years, into Congress. When John Dingell Jr. retired in 2015 as the longest-serving member in the House, he was the forty-third dean of the U.S. House of Representatives, having served in Congress for sixty years (1955–2015). During the early 1970s, Dingell was the chairman of the House Subcommittee on Fisheries and Wildlife Conservation and the Environment of the Committee on Merchant Marine and Fisheries during the enactment of the 1973 Endangered Species Act (ESA). He is the celebrated congressional champion of the ESA, authoring the final version that became law and engineering it through Congress as well as the earlier versions of the law enacted in 1966 and 1969. *Collection of the U.S. House of Representatives*

But Dingell didn't trust that every state would be compliant. Defensive as he was of sportsmen and their financial contribution to conservation, Dingell worried that state-level decisions would be self-serving, guided by the parochial interests of sportsmen that controlled the subculture of the state fish and game commissions and agencies.[6] Sportsmen's annual hunting and fishing license fees funded the state wildlife management agencies, and so huntable game species were the primary focus of state agencies. A uniform, coherent national policy was required to protect endangered species, Dingell believed, not "a series of unconnected and disorganized policies and programs by various states," which would compound confusion.[7]

However, it was recognized that the states were better equipped for day-to-day management and enforcement given their large numbers of biologists and enforcement officers. It seemed that the best balance would be achieved if the secretary of the interior had ultimate responsibility and authority to identify endangered species and regulatory jurisdiction. This duty would be shared with the states provided that each entered into a cooperative management agreement that allowed concurrent jurisdiction with the federal government within a framework of federal rules and standards. These would include financial assistance and incentives to assist in the operation of the program, including land acquisition, expected not to exceed $10 million per year. The costs were to be shared (two-thirds by the federal government and one-third by the states) and cooperative agreements reviewed annually. If at any time a state failed to perform, the secretary could immediately intervene and terminate the agreement.

Dingell's H.R. 37 ultimately provided that the federal government "shall provide enforcement" of the act to the states, concurrent with federal enforcement authority. National oversight and uniform control by the federal government were required, with states free to enact stronger laws than the federal government's. Implementation of this provision was slow, and by 1978 only seventeen states had entered into agreements under it.[8] In a lengthy interview I conducted forty-three years later

(2016) in Dearborn, Michigan, John Dingell expressly confirmed his position in 1973 that the states did not have the capacity to administer the proposed law and that their interpretation and enforcement would result in a hodgepodge of disorganized approaches and policies.[9]

Annual funding to the states to carry out their ESA responsibilities has never covered the costs of this mandate. In 2015, a blue-ribbon panel organized by the Association of Fish and Wildlife Agencies, representing all fifty states, calculated the shortfall to be $1.3 billion per year. States' administrative personnel, scientific expertise, and resources are redirected to the goals of the ESA, and yet those goals often go unmet due to a lack of funding and the slow processes of the federal bureaucracy. In fact, situations have arisen where funding originally allocated to recovery activities has had to be redirected to address administrative overhead and legal costs associated with the ESA.[10]

What funding is available takes the form of conditional grants-in-aid, which can create often irresistible incentives for states to adopt federal policy preferences. To qualify for and retain their federal funding, state governments are required to submit to an exacting federal planning process and then carefully account for each federal dollar spent.[11]

Further constricting states is one of the statutory factors under the ESA for considering the status of a species, equally applicable to listing and to delisting: "the inadequacy of existing regulatory mechanisms." These mechanisms are not defined by the ESA or its implementing regulations and are instead evaluated (and often litigated) on a case-by-case basis. Programs established by state law are more likely to be considered "regulatory" in nature, while programs carried out under gubernatorial executive order—and thus more easily modified or canceled—are less likely to be considered "regulatory." Thus, if regulatory mechanisms are *inadequate* to protect a species, it must be listed. If they are *adequate*, it need not be listed.

THE GREATER SAGE-GROUSE: AN UNPRECEDENTED CONSERVATION EFFORT

Widely recognized as the iconic species of the "sagebrush sea," the greater sage-grouse was first described by Meriwether Lewis in a journal entry dated May 2, 1806, noting that the species is "invariably found in the plains."[12] A large (twenty-two to twenty-nine inches long), nonmigratory, gallinaceous bird with a range that sprawls over 165 million acres in eleven states (California, Colorado, Idaho, Montana, Nevada, North Dakota, Oregon, South Dakota, Utah, Washington, and Wyoming), it is also an ecological indicator species for 353 species that share the same ecosystem, including the pygmy rabbit, pronghorn antelope, badger, loggerhead shrike, mule deer, golden eagle, western burrowing owl, greater short-horned lizard, and sage thrasher.[13]

The greater sage-grouse's range has declined 44 percent from its historic range of 297 million acres. It is a sagebrush-obligate species, feeding exclusively on sagebrush in the winter and relying on sagebrush for cover from predators during nesting and brood-rearing seasons. The population of greater sage-grouse is thought to have once been as high as 16 million, but as settlers spread across the plains through the nineteenth century, fragmenting the landscape, the species declined, and by the early 1900s perhaps 1 million sage-grouse remained.[14] Legendary conservationist Dr. William T. Hornaday organized a campaign against overhunting of sage-grouse, publishing a 1916 pamphlet titled "Save the Sage Grouse from Extinction: A Demand from Civilization to the Western States," which was the first published prediction of the species' possible extinction.[15] Awareness of the sage-grouse's decline gradually spread, and in 1954 the Western Association of State Fish and Wildlife Agencies (WAFWA) established the Sage and Columbian Sharp-Tailed Grouse Technical Committee to study sage-grouse and make recommendations.[16]

By the 1990s, the primary threats to the species included habitat fragmentation due to land subdivision, exurban development, oil and gas development, wildfires, and the spread of invasive species and conifers.[17] In 1994, alarms were sounded regarding a possible ESA listing of the greater sage-grouse, and conservation efforts grew. By 1995, every state in the sage-grouse's range had launched a conservation program to benefit the species.

Greater sage-grouse (*Centrocercus urophasianus*) are known for their elaborate mating rituals. Here, a male spreads out his tail feathers, puffs up his chest, and uses air sacs to emit a distinctive booming sound to attract mates. In an effort to avoid the species' listing under the Endangered Species Act, diverse stakeholders banded together to create the largest voluntary program ever established for the benefit of a single species. *Tom Reichner/Shutterstock*

The species was formally petitioned for listing under the ESA in 2002 and was granted "candidate" status. In 2004, WAFWA produced a "Rangewide Conservation Assessment of Greater Sage-Grouse and Sagebrush Habitats,"[18] followed in 2006 by the "Greater-Sage Grouse Comprehensive Conservation Strategy,"[19] a document that laid out more than 300 conservation activities for federal agencies, state governments, local governments, and NGOs.[20] In 2008, it entered into a memorandum of understanding with six federal agencies (the BLM, FWS, Forest Service, Natural Resources Conservation Service [NRCS], U.S. Geological Survey, and Farm Service Agency) to better facilitate sage-grouse conservation.[21]

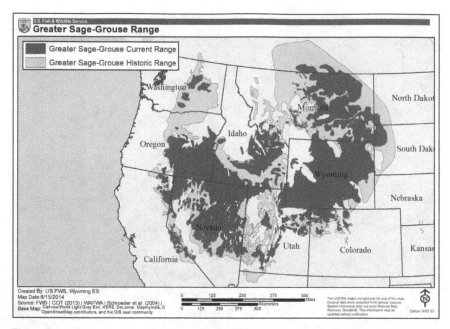

The range of the greater sage-grouse has declined 44 percent from its historic 296,646,000 acres to just 165,168,000 acres. *Courtesy of the U.S. Fish and Wildlife Service*

Hoping to head off the petition to list the greater sage-grouse as threatened, Wyoming Governor Matthew Mead (a Republican) and Colorado Governor John Hickenlooper (a Democrat) proposed a state–federal collaboration in 2011. The Obama administration's secretary of the interior, Ken Salazar, partnered with eleven states in the sage-grouse's range, all four federal land management agencies, industry, local conservation organizations, and landowners, ultimately evolving into the Sage Grouse Taskforce.[22] Over half of the species' range was managed under Bureau of Land Management Resource Management Plan amendments adopted in order to protect the sage-grouse.

Diverse stakeholders were involved in greater sage-grouse conservation, including federal land management agencies, public–private collaboratives, and states. Every one of the eleven states in the range developed a state plan for sage-grouse conservation using a variety of tools, including legislation, executive orders, and other regulations. Altogether, more than $1.5 billion was spent on sage-grouse conservation between

2005 and 2020, by far the largest voluntary program ever established for the benefit of a single species.[23]

The impact of greater sage-grouse conservation efforts, including public–private partnerships, was readily apparent on the ground. For example, in Harney County, Oregon, local ranchers took the initiative to develop a CCAA to benefit greater sage-grouse. The Harney County Stock Growers Association invited the Harney Soil and Water Conservation District (HSWCD) to give a presentation on CCAAs at its 2011 annual meeting,[24] leading to the formation of a committee that worked with the FWS to develop a programmatic CCAA to benefit greater sage-grouse.[25] The Harney County Greater Sage-Grouse CCAA, completed on April 25, 2014, recognizes that properly managed ranching operations can benefit greater sage-grouse and provides that participating landowners will be protected from take prohibitions under the ESA should the greater sage-grouse be listed.[26]

Oil and gas development is one of the primary threats facing the greater sage-grouse. The Jonah Field in Wyoming is the sixth-largest natural gas field in the United States, and it sits directly on top of one of the wildest and most expansive remaining stretches of greater sage-grouse habitat. The land seen here has already been cleared of sagebrush and soon will be crisscrossed with roads, drilling rigs, and buildings that disturb sage-grouse and cause them to abandon their traditional leks. Directional drilling (which reduces the surface occupancy of a given oil and gas development) offers the possibility of reducing habitat disturbances associated with the industry, albeit at greater expense. *Photograph by Noppadol Paothong*

Landowners participating in the CCAA must implement specified conservation measures. All landowners must agree not to allow further fragmentation of sage-grouse habitat on their land, and landowners must also adopt other conservation measures appropriate for their lands, to be determined as each landowner is brought into the program.[27] Conservation measures are drawn from the Oregon Department of Fish and Wildlife's sage-grouse conservation strategy and include measures such as prescribed fire, invasive juniper removal, artificial water sources for livestock, and many more.

HSWCD and the ranchers of Harney County have been widely recognized for the partnership that led to this CCAA. The Public Lands Foundation recognized HSWCD with its 2014 Landscape Stewardship Award, and FWS Director Dan Ashe cited it as an example of successful public-private collaboration.[28] Through partnerships—and planning—the greater sage-grouse was conserved, and, in 2015, the FWS announced that the species did *not* warrant listing under the ESA.[29] In the words of local rancher Tom Sharp, "What's good for the bird is good for the herd."[30]

An example of a conservation-reliant species that was recently delisted is the Kirtland's warbler. A migratory songbird, the warbler primarily breeds in young jack pine forests in the Lower Peninsula of Michigan. It is facing two persistent threats—habitat loss due to fire suppression and industrial tree farms and predation by the brown-headed cowbird, a brood parasite that lays its eggs in the nest of another species, which then raises its young. In the 1950s, the warbler's population dropped to just 1,000 birds, and it was listed with the very first group of species in 1967. At that time, the state of Michigan had already been operating dedicated warbler management areas for a decade.

Conservation efforts on behalf of the warbler have been carried out by the Michigan Department of Natural Resources, the U.S. Forest Service, the FWS, the Wisconsin Department of Natural Resources, the U.S. Department of Agriculture (USDA) Wildlife Services, the Nature Conservancy, Huron Pines, the American Bird Conservancy, and others, including private forest landowners. As a result of this program, the FWS was able to conclude that active management of both cowbirds and forest habitat for the Kirtland's warbler met four factors for delisting despite its conservation-reliant nature: "a conservation partnership capable of continued management, a conservation plan, appropriate binding agreements (such as memoranda of agreement) in place, and sufficient funding to continue conservation actions into the future."[31]

The species was delisted in 2019, at which time it had recovered to more than 2,000 breeding pairs, with more than 127,000 acres being managed for the warbler.[32] State, federal, and private conservation efforts for the species have continued unabated since delisting.

The Kirtland's warbler (*Setophaga kirtlandii*) was listed as an endangered species from 1967 until 2019. A conservation-reliant species, the warbler depends on early successional jack pine forests for breeding habitat. It was successfully delisted due to collaboration and innovative forest management carried out by the state of Michigan, private landowners, and the U.S. Forest Service. *Matthew Jolley/iStock by Getty Images*

Conservation programs for the Kirtland's warbler include brown-headed cowbird control through trapping and rotational management of forests across the species' range. Historically, fire would periodically thin out jack pine forests, and as the forest rejuvenated itself, young stands suitable for warblers would appear. Today, forest stands are instead clear-cut, replanted in jack pine, allowed to mature to commercial viability, and then cut again. By maintaining sufficient forest in this management regime and rotating when and where it is cut, managers can ensure that in any given year, there is enough breeding habitat for the migratory warblers. This approach also allows warbler habitat to be commercially productive, which both generates revenue for warbler management on public land and incentivizes private landowners to participate in the program.

IMPACTS ON PRIVATE LANDOWNERS

Private landowners are subject to the ESA's prohibitions against take, and they may be affected by modifications to designated critical habitat as well. But they can't be liable for impacts to endangered species if there are no endangered species on their lands. So for many years, there was a quiet understanding that if any endangered species on their land just happened to disappear, ideally before the FWS knew of its existence, they would be much better off (the "shoot, shovel, and shut up" approach to protected species).

Regulatory flexibilities for landowners have addressed this perverse incentive. Special rules for threatened species often include broad exceptions for traditional agricultural and forestry practices, while wildlife conservation funding under programs such as the federal Farm Bill often includes a consultation that similarly inoculates landowners from regulation—or prosecution—under the ESA. Today, on balance, the ESA's actual impact on landowners is quite small. In fact, a private landowner with endangered species on their property is now more likely to attract eager conservation partners and funding opportunities than regulators and law enforcement officers.

Foremost among these opportunities are programs run by the NRCS and the Farm Service Agency under the federal Farm Bill. Passed by Congress every five years, the current Farm Bill provides almost $6 billion each year for a variety of conservation programs. These have a veritable alphabet soup of names, such as the Conservation Reserve Program (CRP), Conservation Stewardship Program (CSP), Environmental Quality Incentives Program (EQIP), Agricultural Conservation Easement Program (ACEP), Healthy Forests Reserve Program (HFRP), Regional Conservation Partnership Program (RCPP), and Voluntary Public Access and Habitat Incentive Program (VPA-HIP). All of them incorporate some combination of cost-sharing payments, incentive payments, rental payments, conservation easements, and other mechanisms to fund farmers, ranchers, foresters, and other landowners willing to carry out conservation on their lands.

Some programs merely cover costs. Others pay landowners a market rate and allow conservation to be *profitable*. "Conservation" in this usage is broadly defined. Farm Bill programs vary in their objectives, but all focus on some combination of air quality, water quality, soil retention and quality, wetland restoration, and wildlife. Only the HFRP includes conserving ESA-protected species *specifically* as one of its priority conservation objectives, although all Farm Bill programs benefit wildlife, including endangered species.[33]

Since 2012, a creative regulatory approach has helped prioritize threatened and endangered species within Farm Bill conservation programs. The NRCS of the USDA, together with the FWS, created "Working Lands for Wildlife." Under this program, the NRCS prioritizes conservation funding for a specific, targeted group of ESA-listed or at-risk species. For each species, the NRCS and the FWS conduct a consultation for the species. Thereafter, when the NRCS enrolls individual landowners into its program and funds the landowners' conservation efforts, the *landowners* are covered by the biological opinion issued to the NRCS. Program-endorsed conservation practices are protected from any additional regulation under the ESA, even if the targeted species is subsequently listed or uplisted.[34]

Working Lands for Wildlife has been a resounding success in its first decade. Today, it incorporates eight national and fourteen state-identified priority species and habitats, including the greater sage-grouse, lesser prairie chicken, gopher tortoise, golden-winged warbler, southwestern willow flycatcher, and monarch butterfly. It works in forty-eight states and has invested hundreds of millions of dollars to engage more than 8,400 landowners in conservation of more than 12 million acres. In addition, it has advanced our scientific understanding of both the threatened, endangered, and at-risk species it specifically addresses and the landscapes, ecosystems, and biomes that they call home.[35]

Through this program, at last, there is a mechanism for landowners to work on endangered species habitat and not only avoid prosecution for take (provided by HCPs with no surprises) and future regulation or obligations under the ESA (provided by SHAs and CCAAs) but also in fact be *financially compensated* for the value of the conservation they provide for endangered species. It may have taken almost forty years (1973–2012), but, at last, a true incentive for private endangered species conservation has been applied—though thanks to the Farm Bill and not the ESA.

Working Lands for Wildlife is unique for the *regulatory assurances* it provides landowners. Plain assistance for conservation of fish and wildlife, including threatened, endangered, and at-risk species, also comes from other Farm Bill programs and from the FWS's own Partners for Fish and Wildlife program. Partners for Fish and Wildlife has a separate, congressionally authorized budget of $50 million per year. It has engaged more than 50,000 landowners and more than 6,600 partner groups, leading to the voluntary restoration of more than 6.4 million acres of land, including more than 4.5 million acres of upland habitat, more than 1.5 million wetland acres, and more than 13,500 miles of stream habitat.[36] The USDA also maintains staffed service centers in almost every county in the nation where landowners can go for assistance in conservation planning and accessing technical and financial assistance.

States also work with private landowners, including through both their fish and wildlife agencies and their forestry agencies, to provide technical assistance and conservation planning. In many cases, states also offer their own landowner assistance programs, such as habitat incentives, migration corridor and riparian area protection, pollinator or game species food plots, hunter public access, seed distribution, and more.

Finally, countless NGOs and land trusts partner with private landowners in numerous ways. These programs provide material and labor for conservation work on private lands and may also include financial assistance for landowners.

COMPENSATORY MITIGATION

A final regulatory flexibility for landowners under the ESA is still maturing today but may ultimately prove more important than any other: compensatory mitigation. Compensatory mitigation, which has been practiced for decades for *wetlands* conservation but is still relatively novel for *wildlife* conservation, has the potential to make threatened and endangered species habitat a financial asset that can be quantified, bought, and sold in free markets, harnessing the creative power of capitalism for threatened and endangered species conservation.

"Mitigation" means avoiding, minimizing, and compensating for impacts. "Compensatory mitigation," the last step of the three, refers specifically to "compensating for an impact by replacing or providing substitute resources or environments." Under the ESA, compensatory mitigation may be required, and it may include any of permittee-responsible mitigation, in which conservation is done by the same entity that causes the impacts; in-lieu fee compensatory mitigation, in which the entity causing the impacts pays for a third party to carry out conservation; or mitigation banking, in which the entity causing the impacts purchases credits from a third party that has *already* carried out conservation when it created the bank.[37]

In wetlands conservation, compensatory mitigation is a mature, well-understood, and widely used conservation tool. Wetlands mitigation is governed by regulations issued in 2008 by the U.S. Army Corps of Engineers and the EPA.[38]

Wildlife compensatory mitigation is more complex than wetlands compensatory mitigation. Looking at two equivalent functioning wetlands within the same watershed is virtually an apples-to-apples comparison, and using the simple measure of acres of wetlands is often sufficient. There are many more variables between two parcels of land for wildlife compensatory mitigation. These include the presence of predator or prey species, elevation, climate, moisture, soil conditions, plant communities, and/or human disturbances, all of which affect whether a piece of land is suitable habitat for a species.[39]

In 2016, the Obama administration set out to standardize the FWS's approach to compensatory mitigation, including its role in threatened and endangered species conservation, through new regulations. It was "the first comprehensive treatment of compensatory mitigation under authority of the ESA to be issued by [the FWS]." It followed a series of interagency policies on in-lieu fee programs, wetland mitigation, conservation banking, and recovery crediting systems that had been issued over several years. It was intended to clarify "[FWS] expectations regarding all compensatory mitigation mechanisms recommended or supported by [the FWS] when implementing the ESA, including, but not limited to, conservation banks, in-lieu fee programs, habitat credit exchanges, and permittee-responsible mitigation." The policy was not a major departure from prior practices but reflected lessons learned from more than forty years of implementing the ESA and was intended to "achieve greater consistency, predictability, and transparency in implementation of the ESA."[40]

Unfortunately, the new Obama policy was withdrawn by the Trump administration in 2018.[41] In fact, the Trump administration revoked or withdrew all of the Obama administration's various presidential memoranda, secretarial orders, policies, rulemakings, and other documents relating to mitigation, all within a year and a half of taking office.[42]

In reversing the Obama approach, the Trump administration made it more difficult to develop and implement appropriate and cost-effective compensatory mitigation measures. It also became harder for landowners to identify, develop, implement, and be rewarded for conservation actions.[43] On May 15, 2023, the Biden administration reissued the Obama-era mitigation policies in largely the same form, with some small changes to comply with Supreme Court precedents and minimize controversy.[44]

Compensatory mitigation for unavoidable impacts to wildlife habitat is growing in use. This image shows a successful wetland mitigation project in Cuyahoga Valley National Park, Ohio, one year after completion. In 1995, implementing agencies such as the U.S. Army Corps of Engineers and the Environmental Protection Agency began to permit the replacement of wetlands damaged or lost to development or other human impacts with equivalent functioning wetlands in the same watershed. *Courtesy of EnviroScience, Inc.*

CONSERVATION BANKING

But creative and innovative progress is never far out of reach if we will just look outward and grasp it. One exciting recent development is that in the 2021 National Defense Authorization Act, Congress included a legislative rider directing the FWS to issue a regulation defining conservation banking, which was duly issued in 2022.[45] A conservation bank is a property managed to sustain species either listed under the ESA or otherwise considered at risk. Up to this point, conservation banks have been governed only by a 2003 guidance.[46]

The banker, having invested in the land and funded an endowment to pay for management and monitoring, recoups their investment by selling credits to developers impacting habitat elsewhere. A conservation bank is a free-market mechanism that treats land providing habitat for listed species as a financial asset rather than a liability, paying landowners a market rate for the use of their property to conserve wildlife. This approach allows conservation banks to compete with developers in the free market, and many conservation banks own land that would otherwise have been developed and lost its value as wildlife habitat.

Conservation banks are some of the most effective forms of compensatory mitigation for species because they conserve habitat *before* impacts occur rather than after, are subject to rigorous requirements and oversight, and are invariably well funded. For landowners, they offer some of the highest possible returns on the value of their land, offering the prospect of *profiting* from conservation as opposed to merely *enduring* conservation. They answer Aldo Leopold's charge, that "conservation will ultimately boil down to rewarding the private landowner who conserves the public interest,"[47] better than anything else.

CHAPTER 8

Securing Our Future

Funding Wildlife Conservation

CONSERVING HABITATS *BEFORE* ADVERSE IMPACTS OCCUR AND INCENtivizing individuals to care about the planet's health so that they act in the public interest—these are our challenges for the next fifty years and beyond.

Now more than fifty years old, the ESA is intended to function as the emergency room in a broad-ranging health care system for species, handling the most dire cases. As with health care systems, wildlife conservation efforts that treat only the most desperate cases, at the last possible moment, are highly inefficient and very expensive on a case-by-case or species-by-species basis.

Acting preemptively to conserve species so that they don't ever need the emergency room requires an up-front investment. Given the number of species in crisis today from humanity's past irresponsible actions, we will need more public money, more government flexibility, and more private-sector buy-in in the coming years.

The single greatest obstacle to addressing the extinction and biodiversity crises is a lack of funding. The $1.3 billion budget shortfall calculated by the Association of Fish and Wildlife Agencies' blue-ribbon panel in 2016 proves that our current system of conservation funding is woefully lacking if we are to save the 13,556 species of fish and wildlife already identified by states as species requiring the greatest conservation efforts.

Legislation that would accomplish the panel's goal, called the Recovering America's Wildlife Act (RAWA), would provide the needed $1.3 billion per year in *dedicated*—as opposed to *appropriated*—funding for state fish and wildlife agencies, meaning that it would be guaranteed in each year regardless of other congressional actions. RAWA would provide an additional $100 million to tribal governments for a total annual investment of $1.4 billion.

RAWA was first introduced in the House of Representatives in 2016 and finally passed in 2020, but it stalled in the Senate. In 2021, the legislation was reintroduced, cosponsored in the House by Representatives Debbie Dingell (D-MI), Jeff Fortenberry (R-NE), and eight others (H.R. 2773), and introduced in the Senate by Roy Blunt (R-MO) and Martin Heinrich (D-NM) (S. 2372). Hearings were held in both chambers in 2021, and the legislation passed the House again in 2022. But its final passage in the Senate was disrupted by concerns about paying for it without a dedicated general appropriation. It was reintroduced in the Senate in 2023.

Passage of RAWA would provide state fish and wildlife agencies with the resources they need to build comprehensive, robust programs for conservation of nongame species. It is particularly important that RAWA funding be guaranteed annually, as states will need to develop new programs and hire new staff. RAWA provides a sustainable path toward a future in which state-led wildlife conservation is more comprehensive and more effective and does more to keep species out of the emergency room—as envisioned by Congress fifty years ago.

Across America, the fish and wildlife agencies of the fifty states form the tip of the spear for wildlife conservation. According to a report published by the Association of Fish and Wildlife Agencies in 2017, state fish and wildlife agencies have a collective annual budget of $5.6 billion and nearly 50,000 employees, including 11,000 biologists and 10,100 law enforcement officers. With these resources, they own, manage, or administer wildlife conservation on more than 464 million acres of land and 167 million acres of lakes, reservoirs, wetlands, and riparian areas.[1]

State fish and wildlife agency funding comes from a wide range of sources. In many states, the most important are fees from selling hunting and fishing licenses as well as tags for hunting specific game species. Additional funding comes from state legislatures, some of which have created innovative mechanisms to provide sustainable funding for wildlife conservation. The Missouri Department of Conservation, for example, gets most of its funding from a sales tax. In 1976, Missouri voters amended their state constitution to implement a 1/8-cent sales tax (i.e., one penny on every eight dollars) for wildlife and forest conservation. The program proved so popular that an additional 1/10-cent sales tax was added in 1984 for soil and water conservation and state parks.

Funding for states, federal wildlife conservation programs, and private lands conservation comes from a wide range of sources. An increasingly important source of funding for states are federal grants under the Federal Aid in Wildlife Restoration Act (popularly known as the Pittman-Robertson Act of 1937), which provides federal funding for wildlife conservation and restoration programs to states on a competitive grant basis. The act established a federal excise tax of 10 percent (amended to 11 percent in 1955) on hunting equipment, such as guns and ammunition. In 1970 and 1972, excise taxes of 12 percent on handguns and archery equipment were added to this revenue stream. The success of this program led to the 1950 Federal Aid in Sport Fish Restoration Act (popularly known as the Dingell-Johnson Act), which funds fish restoration, conservation, and management through an excise tax on fishing equipment of 11 percent. The act was supplemented and expanded in 1984 to include boats, motors, fish finders, marine motor fuel, and so on by the Wallop-Breaux amendment. The combined revenues of these programs provide federal funding for countless projects that expand conservation at the state level, including acquisition of lands for management areas and fish hatcheries, and they have a profound influence on wildlife education, scientific research, and management. In 2023 alone, these programs together provided more than $1.6 billion in funding, and, to date, they have accounted for more than $42.8 billion (in inflation-adjusted 2022 dollars).[2]

Another important resource are State Wildlife Grants (SWGs), established by Congress in 2001 to provide funding for state agencies for "wildlife and their habitat, including species that are not hunted or fished."[3] Since 2002, this program has provided more than $1.128 billion for non–game species conservation by state fish and wildlife agencies.[4] In order to qualify for SWG funding, states must prepare and regularly update State Wildlife Action Plans (SWAPs), which provide a road map for nongame conservation in each state, including identification of "species of greatest conservation need," totaling 13,556 nationwide. SWAPs identify species in need of conservation and aid in prelisting conservation and recovery; their recommendations can be implemented by federal and private conservation groups as well as state fish and wildlife agencies. Finally, grants to state fish and wildlife agencies from the Cooperative Endangered Species Fund under section 6 of the ESA have averaged approximately $50 million to $65 million in recent years,[5] an anemic $1 million to $1.3 million per state (on average).

At the FWS, law enforcement consumes approximately $80 million for permitting, enforcement, and trafficking, including obligations under the ESA, the Migratory Bird Treaty Act, and other laws,[6] plus $40 million for law enforcement on national wildlife refuges.[7] Monitoring of migratory birds has its own line item of $29 million, while monitoring of endangered species is included under the broader recovery function.[8] Partners for Fish and Wildlife has its own funding, authorized by the Partners for Fish and Wildlife Act in 2006,[9] currently set at a little more than $50 million per year.[10]

In each of the past five years, Congress has appropriated approximately $20 million for listing and critical habitat designation, $100 million for planning and consultation, $30 million for conservation and restoration, and $100 million for recovery. Total funding has fluctuated between $240 million and $270 million.[11] Half of the total funding available is for planning activities, not for funding on-the-ground conservation. Moreover, within the various planning functions under the ESA, mandatory, judicially enforceable commitments, such as consultations with federal agencies, inevitably take priority over discretionary activities, such as review and approval of HCPs and CCAAs. In some cases, money even has been transferred from the recovery function to the consultation function.[12] Thus, resources are shifted from some of the highest-value conservation work that the FWS does, partnering with private land-owners, to other obligations that are less valuable for wildlife but more pressing for the agency.

Every one of these budget items would benefit from more funding. But funding alone will not guarantee that the money is spent effectively. Funding intended for recovery and collaborative conservation needs to be protected from diversion to programs such as listing and consultation, especially due to litigation. And the Services need to prioritize training and resource allocation to make sure that their staff across the nation are fully and uniformly equipped to engage private landowners and other stakeholders to encourage voluntary and collaborative conservation efforts.

Numerous other federal programs provide funding for wildlife conservation, including on private land, and all of these should be recognized, celebrated, sustained, and expanded. These include programs administered by the U.S. Forest Service, the BLM, the Department of Defense, and the EPA.[13] Another important source of wildlife conservation grants is the National Fish and Wildlife Foundation (NFWF), which provides funds to conserve wildlife and their habitats, in partnership with corporate America.[14] In 2022, NFWF funded 841 projects across the nation with a total of $414.4 million, which was a typical amount for recent years. These grants were matched by a further $319.8 million from grantees, providing a total conservation impact of $734.2 million.[15]

Both the FWS and the National Oceanic and Atmospheric Administration make grants for coastal conservation programs, which include barrier islands, shorelines, dunes, wetlands, estuaries, water quality, historic and cultural sites, scenic vistas, and public access.[16] The FWS's Small Wetlands Acquisition Program purchases easements, including those on private land, to protect waterfowl habitat, chiefly in the prairie pothole region of Minnesota, Montana, North Dakota, and South Dakota as well as in Idaho, Iowa, Maine, Michigan, Nebraska, and Wisconsin.

Altogether, there are more than 34,000 of these easements on small wetlands covering approximately 4 million acres, including 3.25 million acres of easements on private land, protecting grasslands and wetlands in roughly equal proportion.[17]

Natural disasters impact fish and wildlife resources as well as people, and a number of the programs that have been established to respond to disasters include funding for conservation. Some of them are ongoing programs, such as the Emergency Conservation Program, the Emergency Forest Restoration Program, and the Emergency Watershed Program.[18] In addition, when severe natural disasters strike, Congress often passes one-time disaster relief bills, which can include wildlife resources. For example, following Hurricane Katrina and the other storms of the 2005 hurricane season, Congress passed a disaster relief package that included a supplemental appropriation of $199.8 million for agricultural conservation programs. It also provided $404.1 million for a one-time enrollment of private land into ten-year conservation contracts. Following Hurricane Sandy in 2012, Congress again passed a relief package that included a supplemental appropriation of $218 million for agricultural conservation programs. It also appropriated $829.2 million for the Department of the Interior to repair and rebuild coastal assets and promote coastal resiliency.[19]

In 2021, Congress took a bold step toward recognizing and funding America's extraordinary need for more investment in wildlife conservation. The bipartisan Infrastructure Investment and Jobs Act appropriated $28.1 billion over five years to the Department of the Interior, including $455 million to the FWS. The 2022 Inflation Reduction Act funded numerous other critical conservation efforts, including $20 billion for voluntary conservation on private lands through the USDA, $450 million for private forestland management, $1.5 billion for urban forestry, $2 billion for national forests, $2.6 billion for coastal protection and restoration, $125 million for ESA-listed species recovery (half to be spent on the backlog of species needing recovery plans and half on recovery implementation), and a variety of programs to subsidize and incentivize the development of clean energy sources.[20] These massive investments in America's natural resources are unprecedented in their scope, enabling targeted programs selected for funding to progress rapidly. The FWS, for example, has identified seven focal areas for funding: the Klamath Basin in Oregon and California; the Sagebrush Ecosystem; the Delaware River Basin; Lake Tahoe; the National Fish Passage Program; oil, gas, and mineral management; and the Ecosystem Restoration Program.[21]

It is important that the existing diversity of funding opportunities for fish and wildlife conservation not limit our horizons. There is ample space for new and innovative funding mechanisms to incentivize and enable new conservation initiatives, especially by private landowners. For example, Congress could authorize "pay for success" contracting, making direct payments to landowners who host endangered species on their lands or restore them to those lands. It could also establish a revolving loan fund to sustain investment in conservation or utilize a floor price–support approach to incentivize private capital to put money into programs such as mitigation credits.[22] Mechanisms such as these would be particularly helpful for landowners and other groups interested in HCPs, which are very expensive to develop and implement.[23] Each of these approaches would be especially valuable because—although requiring public investment—they would incentivize and leverage private-sector investment in wildlife conservation, a sphere that has experienced rapid growth in recent years.

The Farm Bill is the largest and most important single source of federal funding for conservation thanks to its focus on privately owned lands. (Seventy-four percent of the land in the lower forty-eight states is privately owned, 50 percent of which is managed as cropland, pastureland, and rangeland, while 30 percent is managed as forestland.)

Proposed by President Franklin D. Roosevelt to address widespread domestic hunger and falling crop prices in the 1930s due to the Great Depression and the Dust Bowl, the Farm Bill has been regularly enacted for nine decades, combining conservation, forest policy, disaster payments, crop insurance, price supports, and nutrition. Reauthorized every five years, it affects every American who grows, consumes, or hunts and fishes for food.[24] The Farm Bill's drafting—which begins two years before its five-year-long authorization expires—is driven primarily by the private sector. Foresters, grain growers, cattlemen, conservationists, poultrymen, and state departments of natural resources develop their lists separately and combine their needs into a single request that the USDA and the House and Senate agriculture committees can easily digest.

The oldest of the modern Farm Bill's conservation programs dates back to 1985, when the Conservation Reserve Program (CRP) was established to incentivize the planting of cover crops for soil retention, as were "sodbuster" and "swampbuster" provisions that barred recipients of Farm Bill commodity payments from destroying native grasslands or wetlands. The 1990 Farm Bill expanded CRP and introduced the Wetlands Reserve Program (today known as Wetland Reserve Easements), creating an incentive to conserve and restore wetlands. Producers enroll unproductive and marginal cropland in CRP and the Wetland Reserve Easement Program, the latter of which is expected to save more than $2 billion over the next ten years in crop insurance subsidies, commodity payments, and disaster assistance.[25]

Under consideration for the next Farm Bill is a similar offset associated with CRP that could save taxpayers more than $15 billion.[26] Enrolling cropland with low or negative profitability in these conservation programs brings savings for farmers and ranchers as well, ultimately maximizing their financial returns. As the nonprofit Ducks Unlimited urges, "Farm the best and conserve the rest."[27] This program is widely credited with responsibility for the fact that waterfowl are the only type of North American birds to show population gains rather than losses over the past four decades.[28]

Farm Bill conservation programs are constantly evolving, expanding, and changing to meet the conservation needs of the moment.[29] This evolution includes not just funding mechanisms but also regulatory mechanisms offering more flexibility for private landowners willing to invest in wildlife conservation. In 2002, the Conservation Security Program (now the Conservation Stewardship Program) was introduced; the Healthy Forests Reserve Program (originally passed as a stand-alone program in 2003) was incorporated into the Farm Bill in 2008; and in 2014, both the Agricultural Conservation Easement Program and the Regional Conservation Partnership Program were born.

The rebound of the wood duck (*Aix sponsa*) and its continued availability for harvest are testaments to the impact of waterfowl conservation programs such as the Wetlands Reserve Program, which is funded by the Farm Bill. The male wood duck's distinctive plumage makes it a perennial favorite among duck hunters. *Jim Cumming/Shutterstock*

There is a sharp disparity in our ability to drive wildlife conservation on private lands in contrast to wetland conservation. Under the Clean Water Act, we have significant, voluntary, incentive-based mechanisms that work well. But under the ESA, we lack similar mechanisms to help listed species. In the early 2000s, there was an opportunity to restore habitat—and hence improve populations—for listed at-risk species associated with forests. The Healthy Forests Reserve Program was created to avoid listing and promote recovery of listed and at-risk species through the restoration and protection of forest ecosystems. Land that optimized carbon sequestration was given priority, and this was long before carbon sequestration became mainstream.[30]

The U.S. population is expected to grow by more than 120 million people within the next thirty years, and 50 million acres of forest will be converted to non-forest uses—about the size of Alabama and Mississippi combined. Given this risk and the dependence of wildlife and humans alike on private lands, the Farm Bill is a vital tool for maintaining the ecosystem services that we require—from sequestering carbon to maintaining safe drinking water supplies.[31]

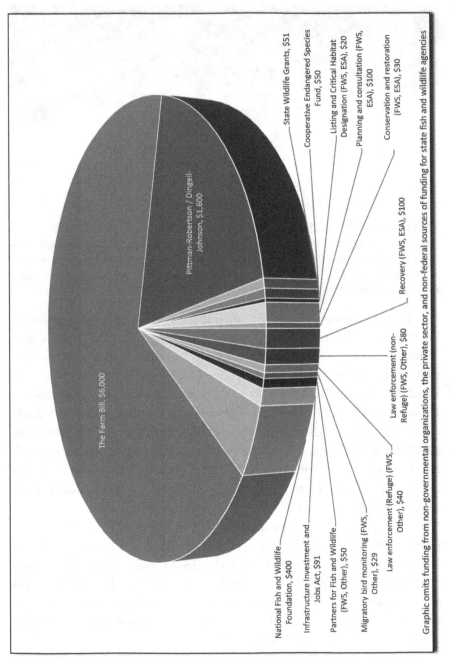

The Farm Bill, $6,000

Pittman-Robertson / Dingell-Johnson, $1,600

National Fish and Wildlife Foundation, $400

Infrastructure Investment and Jobs Act, $91

Partners for Fish and Wildlife (FWS, Other), $50

Migratory bird monitoring (FWS, Other), $29

Law enforcement (Refuge) (FWS, Other), $40

Law enforcement (non-Refuge) (FWS, Other), $80

Recovery (FWS, ESA), $100

State Wildlife Grants, $51

Cooperative Endangered Species Fund, $50

Listing and Critical Habitat Designation (FWS, ESA), $20

Planning and consultation (FWS, ESA), $100

Conservation and restoration (FWS, ESA), $30

Graphic omits funding from non-governmental organizations, the private sector, and non-federal sources of funding for state fish and wildlife agencies

Annual Funding for Wildlife Conservation (in millions of U.S. dollars)

The importance of Farm Bill funding and state fish and wildlife agency efforts for wildlife conservation cannot be overstated. As the diagram on the previous page shows, the various sources of funding through the ESA—and, indeed, the FWS as a whole—are minimal in comparison. Total contributions from NGOs and the private sector cannot be quantified and thus are unfortunately omitted from the chart, as are nonfederal sources of funding for state fish and wildlife agencies.

CHAPTER 9

Making Conservation Strategies Flexible

ONE OF THE MOST ASTONISHING, LEAST FORESEEN ASPECTS OF THE ESA is its extraordinary flexibility—and today, there is a pressing need to utilize it. The twin crises of extinction and biodiversity loss, both worsened by climate change, continue to grow more dire. In order to make progress toward wildlife conservation and recovery and lessen our dependence on the emergency room of the ESA, we need to use and improve existing flexibilities and develop new ones.

Foremost, we need better implementation of existing regulatory flexibilities. Often, staff and resources are diverted away from high-conservation-value activities, such as approving permits, and redirected to address litigation or other mandatory duties under the ESA. In their haste to meet arbitrary deadlines for species chosen by courts and litigants, the Services may overlook more deserving species or opportunities for more cost-effective conservation. Moreover, variations in funding, staff competence, and policy interpretations between and within different FWS offices can be frustrating for landowners and other stakeholders.[1] More thorough and consistent training, as well as higher staffing levels (ultimately a funding problem), would alleviate this situation.

We should also aim to encourage and implement existing flexibilities faster. Such processes are highly dependent on the availability and expertise of FWS staff and funding levels directed to the appropriate offices. The faster these processes can be completed, the more likely landowners, investors, and others in the private sector will be willing to participate.[2]

We can also make existing opportunities for flexibility *smarter*. For example, expanding opportunities for project proponents (regulated entities) to participate in federal interagency consultations would make them faster and more transparent. A number of regulatory changes adopted in 2019 by the Trump administration focused on this goal, and scholars Timothy Male and Ya-Wei Li of the Environmental Policy Innovation Center found the vast majority of those changes positive.[3] In its 2023 revisions to the 2019 regulations, the Biden administration fortunately left these improvements largely unchanged.[4]

The Biden administration has proposed combining SHAs and CCAAs in order to simplify permitting for private landowners.[5] The proposed mechanism, known as a conservation benefit agreement, would standardize the requirements for voluntary conservation of both candidate and listed species while also simplifying the application processing and expediting application processing. This approach would further increase flexibility for landowners by making it easier to return their property to pre-conservation baseline conditions.

HCPs, another major regulatory flexibility of the ESA, could be improved as well. Through either legislation or regulation, mandatory deadlines or other mechanisms requiring agency action could be implemented to incentivize the development of large-scale HCPs. Faster approvals of HCPs and more effective HCPs could be realized through greater staff capacity at the FWS and/or delegation of supervisory responsibilities for HCPs to state agencies. Earlier habitat acquisition and restoration could also speed the approval process, and a complete record of all extant HCPs (including resources protected) could assist in the development of future HCPs.[6]

New flexibilities will likely arise organically in response to events in the real world, and it is very hard to envision what these might look like. For example, HCPs originated from the unique circumstances confronting the San Bruno Mountain group and unexpectedly transformed the ESA from a law that barred absolutely all incidental private take of listed species to one that allowed human needs and endangered species to coexist. Safe harbor agreements (and then CCAAs) arose from the unique situation of the red-cockaded woodpecker at Fort Bragg, the need for the

U.S. Army to train our soldiers, and the creative brilliance of people like Michael Bean and Robert Bonnie—as well as the courage of Secretary of the Interior Bruce Babbitt and his team in endorsing an unprecedented approach to deconflicting the ESA.

A major factor in recovery of the black-capped vireo (*Vireo atricapilla*) was the establishment of dedicated preserves in its nesting range across central Texas, made possible by a habitat conservation plan—one of the Endangered Species Act's many flexibilities. During the years it was listed (1987–2018), its population grew from 350 known birds to more than 14,000. *AGAMI stock/iStock by Getty Images*

Another idea worth pursuing involves the creative use of tailored regulations for threatened species taken under the ESA. This approach would adopt federal regulations that mimic the proposed state management approach for a threatened species. Thus, state management of the species could be tested and fine-tuned while the species was still listed as threatened, establishing confidence that it would continue to thrive when delisted. This would facilitate use of new and innovative management approaches for threatened species, increasing opportunities for successful conservation of those species.[7]

After a species is listed, designation of critical habitat is mandatory under the ESA. Still, the Services have some flexibility. The ESA states that "the Secretary may exclude any area from critical habitat if he determines that the benefits of such exclusion outweigh the benefits of specifying such area as part of the critical habitat." Conflicts between people and protected wildlife cause distrust and make landowners less willing to engage in voluntary conservation actions or even to host protected species on their lands.

The Services could refrain from designating private land as critical habitat if the designation is likely to cause conflict and if the probability that the designation will benefit the species is low. Examples include situations in which any development is unlikely to have a federal nexus or where habitat modification will not necessarily result in take of the species. A vivid example of this situation is the dusky gopher frog critical habitat designation challenged in the 2018 U.S. Supreme Court case *Weyerhaeuser Co. v. U.S. Fish and Wildlife Service.*[8]

Language in the ESA directs all federal agencies to "utilize their authorities in furtherance of the purposes of [the ESA] by carrying out programs for the conservation of endangered species." This language *authorizes* federal agencies to choose to invest in endangered species conservation, but its potential as an incentive to *encourage* agencies to do so has gone virtually untapped. The FWS and the National Marine Fisheries Services could develop guidance or regulations for agencies to *proactively* develop conservation plans, likely reducing the amount of time spent on eventual consultations. Agencies might even be allowed to modify plans in order to incorporate future actions and offset incidental take caused by them without triggering the requisite consultation process. Some of these changes would require legislative—as opposed to merely regulatory—changes to the ESA, but great potential exists for making the ESA more efficient.[9]

PRELISTING CONSERVATION

Regulatory flexibilities can help incentivize proactive partnerships to conserve threatened, endangered, and at-risk species. Such efforts are often most effective when they are initiated before a species is actually listed, as in the case of the greater sage-grouse discussed in chapter 7. Another recent example of how a flexible, prelisting approach to conservation can work well is the organic process by which the public and private sectors came together after the New England cottontail (*Sylvilagus transitionalis*) became a candidate for listing, ultimately securing the future of a species at risk.

The New England cottontail is a rabbit indigenous to the northeastern United States. Habitat loss, together with its limited geographic territory, high degree of habitat specialization, and increasing competition from the eastern cottontail and the snowshoe hare, has caused the New England cottontail's range and numbers to contract over the past sixty years, and, in 2000, the FWS received a petition to list it under the ESA.[10] In 2004, the FWS determined that the petition presented substantial scientific and commercial information indicating that listing might be warranted and, in 2006, rendered its final decision that listing was warranted but precluded by higher-priority actions, which made the species an official "candidate" under the ESA.[11]

Candidate status for the New England cottontail led to the beginning of a multistate conservation effort involving the FWS and state technical staff. Based on the successes the nascent effort was already realizing, the FWS regional director for the Northeast Region, Marvin Moriarty, concluded that the New England cottontail could be conserved without being listed. An ongoing project with an executive-level oversight group formed[12] with representatives from the FWS and the six states in the cottontail's range as well as the NRCS and an independent NGO, the Wildlife Management Institute (WMI), which provided assistance in coordinating efforts across state lines and managing grants.[13] Each already had a long history of bringing expertise and financial resources to collaborative conservation initiatives, which they then brought to this New England Cottontail Regional Initiative.[14]

After the 2011 multidistrict litigation settlement established a listing deadline for the species in the fall of 2015, there was still time to act on the New England cottontail's conservation, Moriarty realized, prompting him to reach out to the states involved.[15] With a newly imposed firm deadline just five years away, the New England cottontail partners would need to intensify their efforts.[16] By 2012, the year-old group, together with conservation stakeholders, had already developed a framework and structure for all of the work that followed.[17]

The New England cottontail is a vivid example of the benefits of partnership and state agency leadership, as their "prelisting" management of the species prevented it from requiring Endangered Species Act protection. *Steve McDonald/Shutterstock*

The New England cottontail is found only in isolated pockets of New England, and the partners implementing the New England Cottontail Conservation Strategy are working to secure conservation of high-quality habitat and introduce new populations where additional habitat is available. *Courtesy of the U.S. Fish and Wildlife Service*

THE NEW ENGLAND COTTONTAIL CONSERVATION STRATEGY

The New England Cottontail Conservation Strategy, designed to be implemented through 2030, is remarkable for its detail and thoroughness. To develop the Conservation Strategy, the New England Cottontail Technical Committee (also consisting of representatives of the six states, the FWS, NRCS, and WMI) conducted an exhaustive survey of the species' range, incorporating information on potential habitat parcels, habitat models, species occurrence data, aerial photography, land in protected statuses such as under conservation easements, and ongoing habitat management efforts. The centerpiece of the Conservation Strategy was a habitat target: to restore 27,000 acres across the species' range, enough land to support 13,500 rabbits.[18]

From the beginning, there was no question that achieving this goal would require an extraordinary collaborative effort, engaging not just the partner agencies but also private landowners, public land managers, NGO partners, and grant funders. Furthermore, although intended to unfold over fifteen to twenty years, the Conservation Strategy would be central to the FWS's listing decision in 2015, just three years in the future.

The most detailed plan is only as good as its implementation, and the New England cottontail effort has featured careful and consistent tracking of its progress and adjustments as needed. Annual performance reports revisit every objective and all accomplishments in each category in each state, documenting progress and remaining needs and scoring each objective as complete, on schedule, improving, delayed, needing improvement, or inactive.

NEW ENGLAND COTTONTAIL CONSERVATION PARTNERS

As the group began to implement the Conservation Strategy, it became apparent that their explicit focus on the cottontail had inadvertently narrowed their coalition by failing to engage individuals, groups, and funders who prioritized the migratory birds, reptiles, and timber that depended on the same land as the cottontail. Realizing this discrepancy and proactively reaching out to engage with these communities, the group attracted more partners who valued the same early successional forest habitat they were trying to create.[19]

By the time of the 2015 ESA listing decision, the New England cottontail conservation movement had grown from the FWS, the states, NRCS, and WMI to include more than 100 partners, including tribes, private industries, NGOs, and individual landowners.[20] Each played a critical role. The FWS, in addition to leading the conservation effort and working with all the partners, mobilized its national wildlife refuges in the region to conserve the cottontail.

One of the challenges of designing and executing the Conservation Strategy was the need to improve the scientific understanding of the species. To do this, a diverse group of local institutions stepped forward and offered their research expertise and funding to aid in the conservation program. Numerous state and federal agencies stepped up with additional support for captive breeding, as did the University of New Hampshire and the University of Rhode Island.[21] The Roger Williams Park Zoo in Providence, Rhode Island, was a critical partner, having already launched captive breeding programs for the cottontail as well as the American burying beetle, the Karner blue butterfly, and the timber rattlesnake.[22] Its New England cottontail breeding program was an immediate success on its launch in 2010, with eleven captive-bred kits born from six founder cottontails weaned and released in the very first year. The program expanded to the Queens Zoo in Flushing Meadows, New York, in 2016.

Outreach to local landowners, communities, and media organizations has been critical. Rabbits are highly charismatic, and the conservation effort leveraged that with initiatives such as community events, seasonal marketing around the Easter holiday, and photo-ops with politicians.[23] For direct contact with local landowners, NRCS provided funding through its Working Lands for Wildlife initiative, which made the rabbit one of its initial target species when it launched in 2012.

Individual private landowners are critical to the overall Conservation Strategy because one of its overarching goals is to establish connectivity between different populations of the cottontail, allowing movement between them and greater resiliency to threats. Such resiliency may prove particularly critical in the coming years if rabbit hemorrhagic disease, an emerging threat in the western United States, eventually spreads to the range of the New England cottontail.[24]

One indispensable source of outreach to private landowners was the local conservation community, including hunting and fishing organizations, which were able to work with landowners on conservation plans and best management practices as well as directly carry out habitat restoration projects. A partial list of NGO partners in the New England Cottontail Regional Initiative includes local Audubon societies, Pheasants Forever/Quail Forever, the American Bird Conservancy, the Quality Deer Management Association, the Ruffed Grouse Society/American Woodcock Society, and the National Wild Turkey Federation.[25]

NEW ENGLAND COTTONTAIL RESULTS
The effect of this conservation effort on the New England cottontail was spectacular, and its impact on the 2015 ESA listing decision was exactly as the partners had hoped. In a determination published on September

15, 2015, the FWS concluded that the cottontail did not warrant listing under the ESA.

The New England cottontail coalition celebrated their success, with Wendi Weber, who succeeded Moriarty as regional director, writing that "Peter Cottontail"—a nod to the work of the writer Thornton W. Burgess, who had lived in Springfield, Massachusetts—had been saved from becoming an endangered species. Weber further celebrated the contributions of the states in the species' range and the broad coalition of conservation partners that had made restoration of the cottontail a success and noted that the young forest habitat used by the rabbit also supports at least sixty-five other species of wildlife as well as opportunities for outdoor recreation and tourism.[26] Thanks to the proactive, prelisting conservation of the New England cottontail, all of these species are now better situated as well.

CHAPTER 10

Partnerships

The Secret to Success

IN 2010, THE CBD FILED ONE OF HISTORY'S LARGEST ESA PETITIONS, seeking the listing of 404 aquatic, riparian, and wetland species in the Southeast.[1] In response, the fifteen states of the Southeast and the federal government launched the Southeast Conservation Adaptation Strategy to promote conservation throughout the region and the Caribbean.[2] This in turn led to a growing region-wide awareness of the need for all stakeholders to work together in partnership, and regular and productive discussions commenced, including outreach to private landowners.

In 2015, a group of five large forest landowners (Resource Management Service, LLC; Rayonier; Plum Creek; Weyerhaeuser; and Hancock Timber Resource Group) proposed that proactive conservation by private landowners could be an important part of successful at-risk species conservation in the Southeast. Inspired by the Louisiana black bear, which had been listed as a threatened species in 1992, the group pledged that the 12.5 million acres they owned or managed would benefit a diversity of wildlife species as sustainable forestry practices were implemented. The forestry industry had taken the initiative to conserve the black bear by protecting its habitat even while using it to produce commercial forestry products, proving that wildlife and forestry could coexist.[3] By 2015, the bear was thriving, and it was actually delisted in 2016.[4]

The Louisiana black bear (*Ursus americanus luteolus*) was delisted thanks to a voluntary public–private partnership and later inspired other conservation efforts in the Southeast. *Clint Turnage/U.S. Fish and Wildlife Service National Digital Library*

This new effort was expanded under the leadership of David Tenny, president and CEO of the National Alliance of Forest Owners (NAFO), along with NAFO members who collectively manage approximately 47 million acres of land nationwide. In partnership with the National Council for Air and Stream Improvement, Inc., they formalized the "Wildlife Conservation Initiative." In 2023, the group signed a first-of-its-kind memorandum with the FWS, reflecting the increasingly enlightened agency culture, which prefers partnerships to a regulatory, command-and-control approach. Under this agreement, the landowners and agencies will continue collaborative conservation efforts, share crucial data on species and habitats, implement conservation actions based on research and the best available science, and communicate with the public about the importance of private lands for wildlife conservation.[5]

As the Wildlife Conservation Initiative was moving toward its 2015 launch, the first stirrings of a complementary shift in agency culture were taking place within the FWS. Then-FWS Southeast Regional

Director Cindy Dohner was wrestling with the CBD's 2010 petition to list 404 species. Dohner saw that the situation called for an innovative, voluntary approach to conserving these species and set out to realize this vision. She soon found a kindred spirit in her colleague Wendi Weber, who, as regional director in the Northeast Region, was spearheading the successful collaborative, prelisting conservation effort that saved the New England cottontail. Both Dohner and Weber became champions for voluntary collaborative conservation within and outside the FWS, and senior agency leaders from the Southeast first conceived the name "Conservation Without Conflict." Dohner and Weber then briefed the FWS Directorate on the Conservation Without Conflict approach, and, concurrent with the growing partnership with private forest owners, the concept was endorsed and implemented across all FWS regions beginning in 2017.[6]

The National Alliance of Forest Owners' Wildlife Conservation Initiative is transforming commercial forest management through a first-of-its-kind memorandum with the U.S. Fish and Wildlife Service. This photograph shows a group of leaders in collaborative conservation during an August 3, 2023, tour of land in Santa Rosa County, Florida, managed by Resource Management Service, LLC. The group includes conservation leaders from the U.S. Fish and Wildlife Service, the Department of Defense, state fish and wildlife agencies, and the forestry community. *Ethan Brietling/National Alliance of Forest Owners*

The movement soon expanded beyond the FWS, as Dohner and Weber joined with a larger group including Jimmy Bullock (senior vice president of forest sustainability at Resource Management Service, LLC) and James Cummins (executive director of Wildlife Mississippi), who shared the idea that voluntary proactive collaborative conservation recognizing the importance of working farms, ranches, and forests could achieve greater conservation success than mandated regulatory approaches. The leadership of this group expanded to include Collin O'Mara (president and CEO of the National Wildlife Federation), Christy Plummer (vice president for conservation of the Theodore Roosevelt Conservation Partnership), David Tenny (president and CEO of NAFO), and Steve Williams (at that time president of the WMI, since retired).

The fledgling group held its first meeting on August 20, 2017, and formally adopted the name "Conservation Without Conflict," describing both the new coalition and the approach it espoused. The roots of Conservation Without Conflict can be traced back to the Louisiana black bear conservation effort in the early 1990s and thereafter the "Wild Goose Chase," which was "an informal meeting of thought leaders from the forestry sector, conservation organizations, and the FWS convened by a conservation leader from Shreveport, Louisiana named Murray Lloyd at Aldo Leopold's 'Shack' outside Baraboo, Wisconsin."[7] These campfire conversations over three years promoted innovative, collaborative approaches to wildlife conservation and created a lifelong network of trusted friends and partners that today, almost thirty years later, continue to work together to further collaborative conservation of species and landscapes across the United States.

In 2017, Conservation Without Conflict held its first meeting, convened by Dohner, Weber, Bullock, and Cummins. That initial 2017 meeting, which attracted just thirty participants in Atlanta, featured a paper by Dohner and Weber that articulated for the first time in writing two central ideas: (1) private landowner economic viability and species conservation need not be mutually exclusive, and (2) voluntary proactive approaches often achieve greater conservation results than ones that are regulatory in nature.[8] A statement of purpose followed, signed by Dohner, Weber, Cummins, and Bullock:

Aldo Leopold (1887–1948), widely recognized as the father of wildlife conservation in America, photographed in front of his shack outside Baraboo, Wisconsin, in 1940. *Courtesy of the Aldo Leopold Foundation and the University of Wisconsin–Madison Archives*

We value land for the recreational experiences it affords, the well-being it provides our families, the support for our way of life, the ecological benefits it delivers to our communities and for the wildlife it sustains. Land and wildlife unite us and enhances our quality of life . . . Conservation without Conflict is both an approach and a coalition. Members have diverse goals and values, including economic profit and sustainability, hunting, fishing, and other outdoor recreation, conservation, national security, and public service, but we all come together around common conservation interests and a sense of good land stewardship. . . . A collaborative approach, along with appropriate and effective incentives that recognize the benefits landowners provide to America's fish and wildlife, can help landowners keep working lands working. This approach realizes the economic and cultural benefits of working landscapes, allows recreational opportunities, and contributes to the enhancement of important habitats. Collaboration that promotes this essential balance of mutual gain among partners is more successful in providing conservation at the scales needed and will be more sustainable than a relationship relying on regulatory approaches alone or at all. This is the essence of Conservation without Conflict.[9]

Dohner, Weber, Bullock, and Cummins organized many of Conservation Without Conflict's formative meetings, and all serve either as advisers or on its Steering Committee to this day. I have worked with them over the years on numerous wildlife and habitat conservation initiatives, and it has been my honor to call all of them my friends and colleagues. Their hard work, selfless dedication, and willingness to put our nation's wildlife ahead of all else are an inspiration to me, and I believe America can learn much from them and from the fruits of their labors. Few know how to move the needle as they each do individually and as a team. They are formidable!

From the beginning, Conservation Without Conflict has maintained its core purpose and vision, embracing the importance of private land for fish and wildlife conservation and the necessity of working collaboratively with landowners to achieve conservation. It embraces incentives for landowners, the best available science, economic and cultural benefits of working landscapes, outdoor recreation (including hunting and fishing), and early, proactive conservation interventions. Conservation Without Conflict recognizes that everyone is better off when species are conserved without listing under the ESA but also that when a species is listed, there is still room for collaboration, regulatory assurances, and broad funding for private lands conservation to conserve and recover species without resorting to litigation and other forms of conflict.[10]

The organization continues to hold regular meetings, and, after three years at its helm (2020–2022), its executive director, Dr. Lauren Ward—who doubled the organization's funding in her first year and attracted the attention of President Biden's White House—was followed in 2023 by Leopoldo "Leo" Miranda-Castro, Cindy Dohner's successor as the FWS's regional director in the Southeast. Under Miranda-Castro, Conservation Without Conflict is identifying best practices for collaborative conservation and building a communications infrastructure and system of regional councils that will allow it to more effectively reach beyond its roots in the eastern United States and influence wildlife conservation across the nation. Undeniably, its influence is growing. President Biden's America the Beautiful Initiative concluded in its 2021 report that "the FWS should enhance support for voluntary conservation efforts by private landowners through initiatives such as Conservation Without Conflict."[11]

Cindy Dohner served as U.S. Fish and Wildlife Service regional director in the Southeast from 2009 to 2017, capping off an extraordinary twenty-five-year career of public service with the agency. She is pictured here holding Louisiana black bear cubs, a species that was successfully recovered and delisted in 2016 during her time as regional director. *Courtesy of Cindy Dohner*

Cindy Dohner, former U.S. Fish and Wildlife Service (FWS) regional director (Southeast) (left), and Wendi Weber, FWS regional director (Northeast) (right), are leaders in the Conservation Without Conflict movement and coalition. They have dedicated years of their careers to promoting collaborative conservation of America's wildlife resources and to achieving a culture change within the FWS to make collaboration a higher priority than regulation. In this photograph, taken in 2012 when both were serving as regional directors, they are shown in Mississippi at the renaming of the Noxubee National Wildlife Refuge as the Sam D. Hamilton Noxubee National Wildlife Refuge in honor of the late FWS director Sam Hamilton, himself a Wild Goose Chase participant and a leader in the growing collaborative conservation movement. *Courtesy of Cindy Dohner*

Jimmy Bullock (right), pictured here on August 3, 2023, in Santa Rosa County, Florida, with U.S. Department of the Interior Assistant Secretary for Fish and Wildlife and Parks Shannon Estenoz (left) and National Alliance of Forest Owners President and CEO David Tenny (center), is another leader of Conservation Without Conflict. His tireless work on behalf of wildlife conservation stretches back decades, as he was the first chair of the Black Bear Conservation Committee that formed in 1990 to lead private, voluntary conservation of that species. Bullock is both a professional wildlife biologist and a forester. Today, he serves as the senior vice president of forest sustainability at Resource Management Service, LLC (RMS), overseeing sustainable forestry and environmental policy and programs and advocacy on forestry issues for RMS-managed timberlands in the United States. He also has responsibility for forest certification and audit programs and leads environmental, social, and governance initiatives for RMS-managed timberlands globally. He serves on numerous boards and conservation organizations. *Courtesy of Jimmy Bullock*

James Cummins, executive director of Wildlife Mississippi, is currently serving as the thirty-sixth president of the Boone and Crockett Club, America's oldest conservation organization. A native of Greenville, Mississippi, and a trained wildlife biologist, Cummins has extensive experience with federal wildlife policy, especially the Farm Bill, beginning with his days working for Senator Thad Cochran (R-MS) and continuing with the Boone and Crockett Club, the Congressional Sportsmen's Foundation, and Wildlife Mississippi. He is a leader in innovating opportunities for private landowners to partner with finance and industry, such as through conservation banking and carbon credit trading, in order to expand opportunities for wildlife conservation on private lands. He was one of the first leaders of Conservation Without Conflict and remains involved in the organization today. *Courtesy of the Boone and Crockett Club*

The key to forming more partnerships with private landowners is providing them with appropriate incentives. When a species is listed or about to be listed and regulatory burdens loom, it may be enough to elicit landowner participation in conservation efforts merely by providing regulatory assurances through ESA programs such as HCPs, SHAs, or CCAAs. But far greater potential gains can be made with financial incentives that can put money into landowners' pockets in exchange for their work on wildlife conservation.

One example of this approach is the Nature Conservancy's "Water Funds," which enable payments from downstream urban water users to upstream rural water users to promote conservation of their shared watershed.[12] This partnership also provides funding and utilizes flexibility to make urban populations more aware of their dependence on natural resources and more connected to other communities, rewards private landowners for conservation, and improves our shared ecosystems and all of their constituent parts.

Water users paying landowners for water quality is an example of how a new idea may arise, making perfect sense in the context of existing conservation programs, initiatives, and mechanisms. We can use partnerships to take advantage of state fish and wildlife agency expertise, the U.S. Geological Survey, and university researchers and other science-based organizations in evaluating species for ESA listing.[13] We can also make use of the connections between species recovery and landscape-scale collaborative conservation. Many species cannot recover without landscape-scale connected habitats, and many landscape collaboratives in turn gain momentum and value for working lands when they address ESA-listed and -candidate species.[14]

Just as there are countless programs that provide incentives, there are numerous mechanisms—various combinations of direct payments and tax incentives—that incentivize private landowner conservation. These can take the form of grants, cost-sharing payments, incentive payments, rental contracts, and conservation easements. In addition to the $6-billion-per-year federal Farm Bill, numerous federal and state agencies, as well as NGOs, operate grant programs that can provide the funding needed to incentivize conservation. Free-market regulatory mechanisms such as conservation banks can be replicated and broadened. A conceptional extension of conservation banks are "habitat exchanges," which have been used for species such as the greater sage-grouse, monarch butterfly, lesser prairie chicken, New England cottontail, Swainson's hawk, and Utah prairie dog. They, too, are market based, but, unlike static conservation banks, they aggregate conservation efforts on a variety of lands, thus increasing flexibility and efficiency and engaging more landowners at lower cost.

The private sector is an increasingly important source of funding for wildlife conservation. Corporations have always been involved in conservation to greater or lesser degrees, especially when required to meet habitat preservation or mitigation standards for projects subject to ESA regulations. For some private organizations, conservation is a profitable business. Resource Management Service, LLC, a timberland investment company that owns and manages forestland on behalf of institutional landowners such as pension funds and trusts, has been a leader in longleaf pine restoration and conservation within the for-profit sector.[15]

In recent years, multiple trends have converged to fuel a new wave of green investments by American businesses. A new emphasis on sustainability pervades everything that companies do, as many consumers today demand consideration of a business's social and environmental impacts. This corporate sustainability movement argues that consideration of these myriad social factors, even at the expense of short-term profits, will change the long-term behavior of corporations in a way that is better for the world, better for humans, better for the corporations themselves[16]—and, of course, better for wildlife and the environment.

In the environmental context, there is a long history of companies investing in conservation as part of their brand. Companies such as Patagonia[17] and Ben and Jerry's[18] have been leaders in putting corporate money into conservation and environmental causes, and many others have followed suit. Annual environmental investments by corporations now exceed $8 billion per year. Much of that money flows to forest, wetland, aquatic, and coastal programs that benefit wildlife, including threatened, endangered, and at-risk species.[19]

Finally, Wall Street is also getting into conservation, making capital available for green companies, including those working on clean energy, electric cars, and other cutting-edge green technologies.[20] Today, there is more corporate money available for investment in conservation than ever before. This presents an opportunity to harness market forces and direct more funding to conservation of wildlife and their habitats. But to do so, the American public will need to think and act strategically—as conservationists, as governments, and as a society—to allocate that money in a way that will have the greatest impact for wildlife conservation.

CHAPTER 11

America Rallies Again

The Promise of the Future

"Cease being intimidated by the argument that a right action is impossible because it does not yield maximum profits, or that a wrong action is to be condoned because it pays."

ALDO LEOPOLD[1]

FIFTY YEARS AFTER THE ESA WAS FIRST ENACTED, AMERICAN WILD-life conservation stands at a crossroads. The extinction crisis that Congress recognized in 1973 has been compounded by a biodiversity crisis and climate change. The need for conservation is more dire than it was fifty years ago.

An estimated 291 species[2] have been saved from extinction by the ESA directly, according to one study. There are still 1,668 species in the United States that are listed as either threatened or endangered, and this number is increasing. In the past ten years, 201 species have been listed, while just forty-six species were delisted due to recovery. More and more species are ending up in the emergency room, at least saving them from extinction. Without ESA protection, many of those species would have become extinct by now.

Unfortunately, the trend toward species extinction shows no sign of slowing. Worldwide, at least 909 species have gone extinct in the past 500 years, and these are merely the species officially declared extinct and listed as such by the IUCN. The IUCN also lists 16,912 species as "vulnerable," 17,344 as "endangered," and 9,760 as "critically endangered."[3] In many cases, we're not even aware of what we're losing. The real numbers are likely closer to the estimates made by ecologists and statisticians such as the late Dr. Edward O. Wilson, ranging from tens of thousands of species to as many as 100,000 species *per year*.[4] This represents an alarming increase in the rate of species loss. Dr. Wilson estimated today's rate of loss to be between 1,000 and 10,000 times that before human intervention.[5] Worldwide, 1 million species are at risk of extinction.[6] According to the World Wildlife Fund (WWF), between 1970 and 2016, the planet suffered a *68 percent decrease* in population sizes of mammals, birds, amphibians, reptiles, and fish.[7] The report, the WWF's latest in a series of biennial papers, garnered global media attention.[8] The WWF concluded, "The findings are clear: Our relationship with nature is broken."[9]

When we imperil our wildlife to the point that it can be saved only through the emergency room that is the ESA, we imperil our own future as well. The question is not *if* we should act now to drastically increase our investments in and commitment to conservation—for we must. The question is *how* and *when* we will act, how we will fund conservation, how we will remain flexible, and how we will fairly share the costs of conservation.

Fortunately, we are a nation of innovators, and we are renowned for our habit of forming voluntary and collaborative groups to solve collective problems. Congress envisioned the ESA as a rigid, top-down, command-and-control law, supported by federal funding for habitat acquisition and state fish and wildlife agency conservation. In practice, within the past thirty years, the ESA has proven highly flexible, encouraging partnerships and investments from all sectors of the American economy and American life.

When we look back over the past fifty years, especially the past thirty years since Secretary of the Interior Bruce Babbitt created regulatory flexibility in applying the ESA, wherever we see wildlife conservation success stories, we see people working in collaborative partnerships. These can be explicit and targeted, the objective of a group like Conservation Without Conflict or the National Alliance of Forest Owners' Wildlife Conservation Initiative, or they can operate behind the scenes, as in Louisiana black bear and greater sage-grouse conservation.

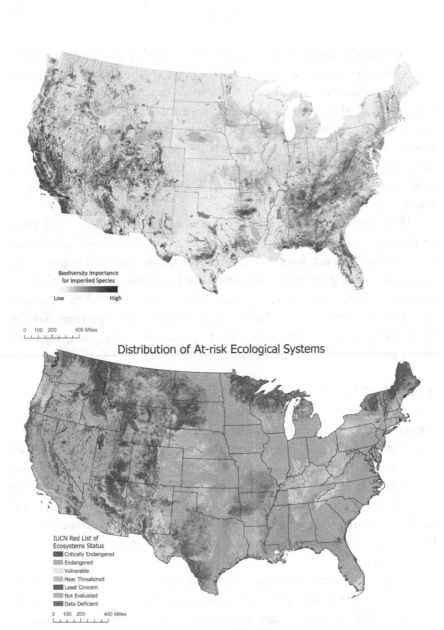

Distribution of At-risk Ecological Systems

Biodiversity Importance for Imperiled Species

Low — High

0 100 200   400 Miles

IUCN Red List of Ecosystems Status
- Critically Endangered
- Endangered
- Vulnerable
- Near Threatened
- Least Concern
- Not Evaluated
- Data Deficient

0 100 200   400 Miles

The United States, like the rest of the world, is facing an unprecedented extinction crisis and loss of biodiversity. These two maps from 2022 show the distribution of critically imperiled species (top) and at-risk ecosystems (bottom). *Maps copyright © 2022, NatureServe, 2550 South Clark Street, Suite 930, Arlington, VA 22202, USA. All Rights Reserved.*

Historically, the burden of conservation funding has not been carried solely by government agencies but has been led by land trusts, sportsmen's organizations, charitable foundations, and private landowners. Often, these other stakeholders have proven more effective than government agencies at resolving conflicts collaboratively and innovating when faced with challenges. One of the challenges for the next fifty years is to continue evolving and innovating and ultimately forge an American conservation movement of which the ESA is just one part—but a highly effective part.

With more funding, more diverse sources of funding, and more diverse programs funded, that vision can become reality. And with trust, mutual respect, interpersonal relationships, sustained engagement, patience, listening, empathy, and established common objectives, the trust necessary for partnerships can grow. In the words of Lesli Allison, executive director of the Western Landowners Alliance, once partners develop trust and a successful track record, they can start "dreaming together" about what else they might accomplish.[10]

THE NEW "VICTORY GARDENS"?

My generation—steeped in the command-and-control mentality— knows that big problems require involvement and investment from government. After World War II, the returning sailors and soldiers turned legislators and executives rushed headfirst into the Green Revolution with the Washington establishment's backing and produced the ESA the way they produced seemingly impossible victories in Europe and the Pacific. But we didn't win World War II without help from civilians. Government participation is a necessary—but not a sufficient—condition for victory.

As just one example, throughout World War II, government agencies, private foundations, businesses, schools, farmers, and seed companies worked together to make sure that both U.S. troops abroad and we civilians at home had enough food. With donated land, instructions, produce, and seeds, individuals and communities grew vegetables and fruits to stretch the food supply. These were known as "victory gardens," and Americans largely avoided hunger as a result.

Europeans, ravaged by the war, were plagued by hunger. To help stockpile grain to send them, President Harry Truman in 1947 asked Americans to give up meat on Tuesdays and poultry and eggs on Thursdays. Americans even volunteered to "starve" in a University of Minnesota study intended to learn the most effective methods for nourishing the war's hunger victims in Europe. Americans have an incredible capacity to sacrifice in order to achieve a goal, and we achieved success while defying steep odds.

Today, "pollinator gardens" might be the new victory gardens in our race against time. Pollinators are bees, butterflies, moths, bats, birds, and other creatures that spread pollen from one plant to another, fertilizing them and allowing them to reproduce. Without them, Earth's biosphere—and our food supply—would rapidly collapse. Today, there is a growing awareness of the dire threats posed to pollinators by habitat destruction, climate change, and other human activities. Schoolchildren and community groups across the nation have long learned about nature by raising one pollinator in particular: the iconic monarch butterfly. Now their efforts are expanding to planting native milkweed to create new habitats for the iconic species, whose population has declined by at least 80 percent from historic levels. The western population of monarch butterflies—which winters in California instead of Mexico like its eastern counterpart—has declined by 90 percent, and its possible extinction due to habitat loss is widely feared—despite a modest population rebound in the past few years.[11] In 2020, the FWS formally announced that the monarch butterfly warrants listing as an endangered species, but a 2014 petition to list it has been precluded by higher-priority listings.[12]

Nearly 25 percent of species of North American wild bees are at risk of extinction, as are a wide variety of non-bee pollinators, including butterflies, moths, flies, wasps, beetles, and ants as well as certain birds and bats.[13] Commercial honeybees are in trouble too. Between April 2022 and April 2023, U.S. beekeepers reported the loss of 48.2 percent of managed honeybee populations.[14] America's pollinators are essential to the production of $15 billion to $18 billion worth of crops each year.[15] In total, insect pollinators contribute $29 billion to U.S. farm income, with almost half of that coming from wild bees and related pollinator insects.[16] Worldwide, the United Nations estimates that pollinators drive approximately $600 billion of the world's economy each year.[17] In addition to pollination services, the natural world and its biodiversity sustain our agricultural crops themselves, providing fish, meat, and other foodstuffs and 50 percent of our drugs and medicines.[18]

The planting and maintenance of gardens for all pollinator species of insects and birds has become a movement all across America, highlighted by First Lady Michelle Obama when she created a pollinator garden at the White House in 2014 and attracted national media attention.[19] The same year, President Obama issued a presidential memorandum, "Creating a Federal Strategy to Promote the Health of Honey Bees and Other Pollinators." This established the Pollinator Health Task Force, which was charged with developing a National Pollinator Health Strategy.[20]

The iconic monarch butterfly (*Danaus plexippus*) has been a candidate for Endangered Species Act listing since 2014, sparking numerous conservation efforts. *Sari ONeal/Shutterstock*

Americans are rallying to face this new challenge. Countless community groups and organizations have followed the Obamas' lead in addressing the monarch butterfly. In 2017, NRCS established the Monarch Butterfly Initiative within the Working Lands for Wildlife program, prioritizing private monarch butterfly conservation for federal Farm Bill funding and offering participating landowners regulatory assurances under the ESA that if the species is listed in the future, they will incur no additional costs or obligations.[21] Numerous Farm Bill funding mechanisms have been used for monarch conservation, including the Environmental Quality Incentives Program and the Regional Conservation Partnership Program.

In 2018, I led the National Capital Area Council of the Boy Scouts of America (BSA) in launching its Milkweed for Monarchs campaign, in partnership with the United States National Arboretum and the National Wildlife Federation (NWF), to plant milkweed and create pollinator gardens throughout the Washington, D.C., area. The BSA has since spread the campaign across the United States with scout troops in every state.

Milkweed for Monarchs is just one of NWF's numerous efforts on behalf of monarchs and other pollinators, as the organization launched a major nationwide monarch conservation campaign in 2015. It became a centerpiece in NWF's children's education curriculum throughout the country. NWF holds an annual Mayors' Monarch Pledge to engage cities and communities in monarch and pollinator conservation. It has reached more than 600 mayors and heads of local and tribal government, engaged more than 6 million people, and restored more than 6,500 acres of monarch habitat.[22] This program complements NWF's long-running Certified Wildlife Habitat program for backyards and gardens and other programs that reach individuals and families in every community across the country.

Industry is also taking a prominent role in monarch butterfly conservation. In 2020, the FWS finalized a CCAA with thirty representatives of the energy and transportation industry in order to conserve monarch habitat along rights-of-way across the nation.[23] This agreement, covering forty-eight states, is the biggest CCAA ever developed. In approving it, the FWS identified 500 other species that also will benefit.

In development for more than two years, the agreement encourages energy companies and highway agencies that control rights-of-way to enroll and manage their land to support monarchs. When fully realized, the plan could cover as many as 26 million acres, an area the size of Kentucky. Enrollees will maintain existing monarch habitat and create new habitat through conservation actions such as "habitat set-asides, timed mowing to avoid peak periods when monarch butterflies are present, use of targeted herbicide applications to selectively treat woody and invasive plants, and other integrated vegetation management best practices that promote beneficial vegetation like milkweed and other flowering plants." In return, should the monarch be listed under the ESA, participants are guaranteed to be protected from further obligations as long as they continue to implement the CCAA, providing them with regulatory and cost certainty as they provide critical infrastructure services to our nation.[24]

Kickoff of the Boy Scouts of America (BSA) Milkweed for Monarchs campaign in 2018. The setting is the National Arboretum, Washington, D.C. Left to right: Les Baron, scout executive for the National Capital Area Council (NCAC BSA); Will Roger, chair, NCAC BSA Conservation Committee; author Lowell Baier; Collin O'Mara, CEO, National Wildlife Federation; Tom Melius, Midwest director, U.S. Fish and Wildlife Service; and Stephen Donnelly, district executive, NCAC BSA. *Photograph courtesy of the National Capital Area Council, Boy Scouts of America*

Governments at all levels are joining together to conserve monarchs. Internationally, the United States, Canada, and Mexico have jointly created the North American Monarch Conservation Plan.[25] At the state level, the Midwest Association of Fish and Wildlife Agencies has taken the lead in developing the comprehensive Mid-America Monarch Conservation Strategy.[26] The strategy lays out conservation actions including restoring, enhancing, and protecting habitat and providing information, education, and conservation engagement opportunities to interested citizens from the plan's establishment in 2018 through 2038. Twenty-nine states are participating, ranging from Texas to Maine. The goal is to establish 1.3 *billion* additional milkweed stems over the next twenty years. Implementation is through state-specific plans carried out in partnership with numerous agencies, other organizations, and private landowners.[27]

In 2019, the Western Association of Fish and Wildlife Agencies followed with its own Western Monarch Butterfly Conservation Plan, a fifty-year (2019–2069) plan to conserve the western population and avoid the need to list the butterfly. The proposed Monarch Action, Recovery, and Conservation of Habitat (MONARCH) Act, currently being considered by Congress, would authorize $62.5 million in federal funds for projects aimed at conserving the western monarch and an additional $62.5 million to implement the Western Monarch Butterfly Conservation Plan.[28]

There is now also a Monarch Joint Venture, modeled after the migratory bird joint ventures established in 1986 by the North American Waterfowl Management Plan, and it has produced its own Monarch Conservation Implementation Plan.[29] The species is also the subject of both a FWS Conservation Efforts Database and a private database, part of the Habitat Conservation Assistance Network, that helps individuals locate local programs and opportunities for monarch conservation.[30]

The wide-ranging diversity of monarch butterfly conservation programs now under way illustrates the potential for broad-based, collaborative, voluntary conservation to transform America. Equally innovative conservation efforts are under way for countless other species, including not only other pollinators such as birds and bees but also the full diversity of plants, animals, insects, reptiles, crustaceans, amphibians, and fish and aquatic species. Together, we can address the biodiversity crisis, prevent extinctions, and begin to heal Earth of the wounds inflicted by our civilization.

HOPE FOR THE FUTURE

When I look at the history of the ESA, I see a road map for our future, a future of more frequent—and more *effective*—wildlife conservation efforts. There is much to learn from our history—too much to fully address in a book such as this one. Listings, controversies, recovery efforts, partnerships, legislation, litigation, regulations, conservationists, environmentalists, sportsmen, landowners, volunteers, and plain Americans all offer lessons learned. Our task now must to be take these lessons and forge them into a broader, more united, and ultimately more effective conservation movement.

The first step must be to make conservation relevant to Americans and once again a part of our national agenda, as it was in the 1960s and 1970s. Today, thanks to the work of countless researchers and advocates, we are finally becoming aware of the depth of the biodiversity and extinction crises. The late Dr. Thomas Lovejoy, who is widely credited with coining the term "biodiversity" in the early 1980s, published extensively on the dangers posed by anthropogenic climate change and deforestation.[31] In this advocacy, he was joined by his contemporary, the late Dr. Edward O. Wilson, who twice won the Pulitzer Prize and was one of the earliest advocates of biodiversity.[32]

Referring to it as the "pauperization" of species, Dr. Wilson linked the biodiversity crisis to two looming disasters: climate change and a worldwide shortage of fresh water. The solutions to all of these problems are similarly linked: by protecting species, we can protect our biodiversity, ecosystems, and genetic diversity, in turn promoting complementary values such as climate resiliency and air and water quality.[33]

Luminaries such as Lovejoy and Wilson have established this vision for us as a goal. Now we need to take control of our future, united by love and concern not only for nature but also for ourselves, each other, and the generations yet to come. The ESA of 1973 is our most powerful vehicle to achieve this end.

Dr. Edward O. Wilson (1929–2021) is known as the "father of biodiversity" in recognition of his groundbreaking research, writing, public speaking, and tireless advocacy for a natural world that he believed can still be saved from destruction. Educated at the University of Alabama and Harvard, Wilson was the world's foremost expert on ants, including both their biology and their behavior. While conducting fieldwork on the islands of Melanesia in the 1960s, Wilson and mathematician Robert MacArthur developed the theory of island biogeography, mathematical equations that defined the species abundance and diversity that could be supported by an island of a given size, now a cornerstone of modern ecology. From this beginning, Wilson expanded his focus to other insects, other animals, and finally humans, becoming a sociobiologist and humanist philosopher. He forcefully argued that humans could coexist with nature and that science, philosophy, and the humanities complemented each other and could all help enrich the human condition. He won a Pulitzer Prize for *On Human Nature* in 1979 and again for *The Ants* in 1991. Late in his life, he returned to the theory of island biogeography and published *Half Earth* in 2016, a call to conserve 95 percent of the planet's biodiversity by preserving intact 50 percent of its ecosystems. This advocacy is now carried on by the E. O. Wilson Biodiversity Foundation, established in Wilson's honor in 2005. Here, Dr. Wilson is pictured with the author at the U.S. Capitol, October 24, 2017. *Photograph by Peter Hershey*

AFTERWORD

In 1962, Rachel Carson wrote in *Silent Spring*,

> We stand now where two roads diverge. But unlike the roads in Robert Frost's familiar poem, they are not equally fair. The road we have long been traveling is deceptively easy, a smooth superhighway on which we progress with great speed, but at its end lies disaster. The other fork of the road—the one less traveled by—offers our last, our only chance to reach a destination that assures the preservation of the earth.[1]

The preservation of Earth is not only in the best interest of our own human species; I believe it is also our moral and ethical responsibility. When Congress debated passage of the ESA back in 1972–1973, the lawmakers saw care for endangered species as necessary to protect humanity from itself and also as an explicit moral duty, lamenting our greed and that "our blind desire for progress enabled us to interrupt the rhythm of nature," that we suffer from "ecological myopia," and that "man's superior intelligence has not yielded him ecological wisdom."

This kind of moral clarity similarly guided the Continental Congress when they adopted the Declaration of Independence, the U.S. Constitution, and the Bill of Rights two centuries earlier. Moral duty formed the United States and grounded our natural laws and the public trust doctrine of jus publicum and emanates from the Judeo-Christian scriptures that are our nation's heritage.

I worry that we've lost our spiritual center and inner compass both individually and as a nation. That inner compass and reprioritizing and refocusing of national interests is what gave rise to the Green Revolution of the 1960s–1970s following World War II. Today, our society is too distracted by the 24/7 news cycle, misinformation, and lust for material wealth and fame to feed our narcissism to care about the scientific and empirical evidence documenting climate change and our worsening biodiversity crisis. Yet now the increasing strength of hurricanes, rapidly spreading wildfires, 100- to 500-year floods, and continuing drought are thrust in front of our eyes, killing both people and thousands of species.

Mollie Beattie, the FWS director from 1993 until her untimely passing in 1996, once stated that "what a country chooses to save is what a country chooses to say about itself."[2] The ESA—Earth's emergency room—will play a critical role in the future, but it is just one tool among many for saving wildlife. There is work that must be done by everyone—by Congress, federal agencies, state agencies, private landowners, industry, NGOs, and private citizens alike. Together, we can recommit ourselves to the vision embraced by Congress when it passed the ESA fifty years ago and make it our goal to conserve endangered species and the ecosystems on which they depend.

To meet this moment, we must all be advocates, working in the communities we inhabit and slowly, as a grassroots movement, building a consensus around wildlife conservation and making it a national goal. We can raise this issue's relevance just as we did during the Green Revolution. In so doing, we will save not only species but also our planet, ourselves, each other, and our children and those future souls in the womb of time.

This is a call to action of the obligation to ourselves, our families, and society for making extinction, biodiversity, and the ESA a national priority again. God has given humans nature's garden, of which we are but a part. It is ours to protect and maintain and not despoil and pollute because it supports our very survival.

Go forth in peace, and embrace the future challenges.

Lowell E. Baier
October 16, 2023
Washington, D.C.

Acknowledgments

This is the story of the Endangered Species Act of 1973: its development over fifty years told in colorful vignettes and stories from a variety of perspectives as people experienced it, including this author. Its history is rich in colorful detail, heartache, agony, and joy. This law is perhaps the most controversial of all environmental laws enacted during the Green Revolution of the 1960s and 1970s, America's period of environmental awakening.

An attempt to capture the history of the Endangered Species Act in a single volume is a daunting task—many of the myriad aspects of the law and its application are worthy of an entire book in their own right. I offer my deepest thanks to the many people who so generously gave their time and expertise to help me sharpen the scope, depth, and organization of this book. The following sages provided invaluable service to help navigate through this weighty subject. In alphabetical order, they include E. U. Curtis "Buff" Bohlen, the late John D. Dingell Jr., H. Dale Hall, Ya-Wei Li, the late Dr. Tom Lovejoy, James R. Lyons, Timothy Male, Dr. John Organ, Frank Potter Jr., the late Nathaniel P. Reed, the late Dr. Lee M. Talbot, Douglas Wheeler, Steven Williams, and David Willms.

Many distinguished and accomplished conservationists, peers, and others generously shared their knowledge, experiences, and counsel in order to improve this book. They offered many invaluable suggestions, and their contributions greatly improved the finished product. I thank them for their time and willingness to be formally interviewed about their organizations, their experiences with endangered species conservation, and their knowledge of the Endangered Species Act. In alphabetical order, they include Lesli Allison, Dr. Ed Arnett, Daniel M. Ashe, Donald Barry, Michael Bean, Bruce Beard, David Belitsky, Mike Brennan,

Curt Brewer, Jimmy Bullock, Dr. Loren Chase, Jean Cole, Ron Crete, James L. Cummins, Dave Cunningham, Cindy Dohner, Ali Duvall, John Emmerich, Steve Forrest, Gary Frazer, Dr. Donald "Pete" Gober, Lynn Greenwalt, Michael Greve, Tim Griffiths, Galon Hall, H. Dale Hall, Dr. Healy Hamilton, Mark Hatfield, Alan and Christine Hogg, Mark Humpert, Jim Inglis, Douglas Inkley, Wendy Jackson, Steve Jester, Lane Kisonak, Larry LaFranchi, Ronald E. Lambertson, Paul Lenzini, James R. Lyons, Timothy Male, Lindell Marsh, Kenneth Mayer, George Meyer, Leopoldo "Leo" Miranda, Miles Moretti, Paul Nickerson, Robert Oakleaf, Dr. John Organ, Chris Parish, Joel Pedersen, Ignacio Jiménez Pérez, Louis Perrotti, Dr. Paul Phifer, Ron Regan, Mark Rey, Brian Rutledge, Dr. John Sanderson, Jen Mock Schaeffer, Dr. Greg Schildwachter, Dwayne Shaw, Dave Smith, Bob Spears, John Spinks, San Stiver, Jim Stone, Scott Talbott, Julie Tripp, Anthony Tur, Tony Wasley, Wendi Weber, Allison Wells, Douglas Wheeler, Nick Wiley, Steve Williams, David Willms, Rich Winkler Jr., and Roger Wolfe. I wish to be clear, however, that all errors and omissions are solely my own. My apologies to those whom I've inadvertently omitted throughout these acknowledgments. Further, I want to thank all the past and current wildlife biologists and administrators at all levels of government that applied and enforced the Endangered Species Act. They are the ones who have written the history of the act over these past fifty years now recorded in this book.

I would be remiss if I failed to acknowledge the giant naturalists, philosophers, and scientists of the past on whose shoulders we stand due to their extensive scholarship on species and extinction. They include Alexander von Humboldt, Dr. Ernst Haeckel, Charles Darwin, George Perkins Marsh, Aldo Leopold, and Dr. Edward O. Wilson.

Next in line is my senior associate, research assistant, and colleague Christopher E. Segal, who joined me thirteen years ago out of Georgetown Law School. I call him "Yoda" because he is invaluable in his depth of research skills, follow-up, diligence, intuitive sense of what's important in evaluating a topic, prioritization thereof, and so on. In manuscript preparation, editing, fact-checking, and attention to accuracy and detail, he is remarkable. This book couldn't have been written without Chris, who did an equal amount of the heavy lifting in deep research and writing. My executive assistant Janice Shoyooee is my daily hero in deciphering legions of handwritten material; anticipating my unrelenting need for perfection, information, and facts; locating lost files; and scrambling to stay ahead of me and keep my schedule organized. Finally, thanks go to Bonnie Baier, who keeps the home fires burning, patiently tolerating my late nights, weekends devoted to writing, exhausting schedule, and publishing deadlines for these past fifty-six years.

This book could not have been written without the oversight and counseling of our New York editor Bernadette Serton. Her invaluable expertise and tireless efforts made this book readable and a contribution to our national dialogue. She brought this author, so used to writing in legal and technical terminology, down to earth for the general public in a colloquial writing style. Moreover, thanks to Jon Sisk, vice president and senior executive editor at my publisher Rowman & Littlefield, who encouraged me to write this book, and publisher Jed Lyons, who recognized the importance of publishing this timely story on the fiftieth anniversary of the Endangered Species Act.

Thanks likewise go to Terry Tempest Williams for writing the foreword. I met Terry in Albuquerque, New Mexico, during the fiftieth anniversary of the Wilderness Act in 2014. Our mutual friend Dr. Douglas Brinkley introduced us. I'd read Terry's work for years and followed her career closely. She's filled the void of environmental consciousness left by Wallace Stegner, Edward Abbey, Rachel Carson, Wendell Berry, and Aldo Leopold, with the deep spirituality of Thomas Merton embodied in his writings. Terry lives in Utah and close to the land and wilderness that she chronicles brilliantly and with elegance. We've connected because we are both grounded in the spirituality Mother Earth provides those who can detach from the clamor and noise of civilization and hear the voices of the Earth and natural world around us. No one is better equipped to write a foreword for this book that is a call to action to protect our fragile landscape from man's actions that pollute and desecrate it, destroying its thousands of species that call it home. I'm indebted to Terry for adding her voice to this clarion call. Finally, my thanks to you the reader for listening to this clarion call. My prayer is that you will embrace it.

Go forth in peace.

Lowell E. Baier

Appendix A

Endangered Species Act Timeline

This timeline summarizes some of the many events in our nation's growing effort to conserve our endangered, threatened, and at-risk animal and plant species.

Pre-1960

1896: In *Geer v. Connecticut*, the U.S. Supreme Court declares that states have the authority to manage their resident wildlife and prohibit the export of game lawfully killed within the state. This case becomes the single strongest precedent and foundational bulwark for a state's sovereign authority to manage its resident wildlife—legally, intellectually, and emotionally. From the states' perspective, wildlife within their borders was theirs to manage at their sole discretion; they believed they actually owned the wildlife. The states' wildlife management systems and organizational structures were built on this premise.

1900: The Lacey Act makes the interstate transportation of game killed in violation of state laws a federal crime. This was the federal government's first foray into wildlife management, which had traditionally been the exclusive domain of state governments.

1903: President Theodore Roosevelt establishes the first National Wildlife Refuge at Pelican Island, Florida, to protect wood storks, brown pelicans, and other dwindling water birds. (Today, national wildlife refuges support nearly 300 endangered and threatened plant and animal species.)

1906: The Antiquities Act authorizes the president to establish by proclamation national monuments on public lands managed by the federal government in order to protect "historic landmarks, historic and prehistoric structures, and other objects of historic or scientific interest." The president is authorized to act independently of any further congressional action. President Theodore Roosevelt uses this authority to protect numerous landscapes that later became national parks, including the Grand Canyon, and every subsequent president has since used this authority.

1914: The passenger pigeon, once the most abundant bird in North America and perhaps the world, becomes extinct, joining a long list of extinct species that include the ivory-billed woodpecker, great auk, heath hen, Carolina parakeet, and others. All of these species were driven to extinction by commercial market hunters overharvesting them for food and/or plumage for fashionable hats and dress.

1916: The United States and Great Britain (on behalf of Canada) sign an international treaty establishing a uniform system of protection for certain species of birds that migrate between the United States and Canada. This is the world's first international treaty for wildlife conservation.

1918: The Migratory Bird Treaty Act (MBTA), implementing the 1916 migratory bird treaty, marks the first explicit preemption of state wildlife management authority by federal law. The earlier Weeks-McLean Act of 1913 had attempted to do this but was found unconstitutional by federal courts. The MBTA, in contrast, is upheld by the U.S. Supreme Court in 1920 in *State of Missouri v. Holland* because it implemented a federal treaty with Great Britain. The U.S. Constitution provides that treaty making is an enumerated power of the federal government, and the Supremacy Clause provides that federal treaties supersede conflicting state laws.

1920: In *State of Missouri v. Holland*, the U.S. Supreme Court upholds the Migratory Bird Treaty Act as a valid exercise of the Treaty Clause of the U.S. Constitution that overrode state ownership and management of migratory birds, including game species.

1928: In *Hunt v. United States*, the U.S. Supreme Court upholds the authority of the federal government to protect federal lands notwithstanding state law, including through management of wildlife, under the power of the Property Clause of the U.S. Constitution. The case arose in the context of Aldo Leopold's famous experiment on the Kaibab Plateau of Arizona, where a program of predator control led to an excessive mule deer population that needed to be artificially reduced. This establishes, for the first time, federal authority over resident wildlife within a single state.

1934: The Migratory Bird Hunting Stamp Act (the "Duck Stamp") placed a $1 tax on every waterfowl hunting license. Today, this has risen to $25. This generates a revenue stream used by the federal government to acquire, expand, maintain, and manage waterfowl refuges, including many areas previously subject to state authority. The expansion of waterfowl refuges creates a much larger footprint of federal land control and wildlife management authority in affected states. Waterfowl management will later develop into a great federal–state collaboration success story in the 1950s, as a series of flyway councils are set up to work cooperatively with states to manage the bird populations sustained by refuges.

1935: Under the leadership of J. N. "Ding" Darling (chief of the Bureau of Biological Survey), the federal government creates a nationwide series of university science courses and PhD research programs housed in land grant universities, known as the Cooperative Fish and Wildlife Research Units program. These ventures are cooperatively managed by the state department of natural resources, the federal government, the Wildlife Management Institute, and the host university. Today, forty-three units exist.

1937: The Federal Aid in Wildlife Restoration Act (Pittman-Robertson Act) directs federal excise taxes on hunting equipment (e.g., guns and ammunition, later expanded to include handguns and archery equipment) into a fund to be distributed to state fish and wildlife agencies according to a formula based on state size and the number of hunting licenses sold. Federal funds must then be matched by state agencies at a ratio of 3:1. This funding, which as of fiscal year 2023 has provided a total of $25.3 billion (in 2022 dollars, adjusted for inflation), is essential to the operation of state fish and wildlife agencies across the United States.

1940: The Bald Eagle Protection Act prohibits the wounding, capturing, molesting, disturbing, or killing of bald eagles and the possession of eagle parts, feathers, nests, or eggs.

1944: The whooping crane population, victim of habitat loss and hunting, reaches its lowest population level, with only twenty-one birds remaining.

1950: Building on the model of the Pittman-Robertson Act, the Federal Aid in Sport Fish Restoration Act (Dingell-Johnson Act) funds fish restoration, conservation, and management through an excise tax on fishing equipment (later expanded to include boats, motors, fish finders, marine motor fuel, and so on) using apportionment formulas and matching funds similar to Pittman-Robertson. As of fiscal year 2023, this program has provided a total of $17.5 billion (in 2022 dollars, adjusted for inflation) to state fish and wildlife agencies.

1960s

1962: Rachel Carson's book *Silent Spring* warns of impacts on wildlife and people from unregulated pesticide use. *Silent Spring* is widely credited with inspiring the environmental movement of the 1960s and 1970s, triggered in part by major environmental disasters, such as acid rain, smog, the burning of the Cuyahoga River (1969), and the Santa Barbara oil spill (1969).

1962: The Bald and Golden Eagle Protection Act amends the Bald Eagle Protection Act to protect golden eagles as well as bald eagles.

1964: First publication of the International Union for Conservation of Nature's "Red List of Threatened Species." The Red List is one of the inspirations for the U.S. Endangered Species Act (ESA), and one of the scientists involved, Dr. Lee Talbot, later plays a key role in writing the ESA of 1973.

1964: The U.S. Fish and Wildlife Service (FWS) launches a captive breeding program for whooping cranes at its facility in Patuxent, Maryland. The first chicks are produced in 1967, and the breeding program is later expanded to numerous other sites, producing cranes for reintroduction into the wild.

1966: The Endangered Species Preservation Act of 1966 authorizes the secretary of the interior to utilize existing authorities, such as the Migratory Bird Conservation Act, the Fish and Wildlife Act of 1956, and the Fish and Wildlife Coordination Act, to acquire land to conserve "selected species of native fish and wildlife."

1967: The first list of seventy-eight endangered species is published by the FWS.

1967: The Tennessee Valley Authority (TVA) begins construction of the Tellico Dam on the Little Tennessee River outside of Knoxville, Tennessee. The dam, designed to raise the hydraulic head at the nearby hydroelectric Fort Loudon Dam, would flood 16,500 acres along a thirty-three-mile reservoir, but the TVA bought and/or condemned 42,999 acres of private land to create housing projects, a model town, and industrial parks around the new lake. The threatened loss of local farms, archaeological sites, burial grounds sacred to Cherokee Indians, and one of the finest trout-fishing streams in the nation galvanized widespread opposition to the project.

1969: The Endangered Species Conservation Act of 1969 expands on the 1966 act, authorizing the compilation of a list of animals "threatened with worldwide extinction," and prohibits their importation without a permit. Crustaceans and mollusks are included for protection along with mammals, fish, birds, and amphibians.

1970s

1970: The National Environmental Policy Act (NEPA) requires that the federal government evaluate the environmental impact of all federal actions "significantly affect[ing] the quality of the human environment." It also establishes the Council on Environmental Quality within the Executive Office of the President to implement the law. NEPA will later be seen as the bedrock of all American environmental law as well as a source of significant delay in federal permitting and major infrastructure projects.

1970: The first Earth Day is celebrated on April 22.

1970: The peregrine falcon and red-cockaded woodpecker are listed as endangered.

1971: The International Conference on the Biology of Whales, held in Virginia, is one of several events that highlight the shortcomings of existing wildlife laws and inspire officials in the Nixon administration to work on crafting new endangered species legislation.

1971: The first lawsuit is filed against the TVA's Tellico Dam project, alleging failure to comply with NEPA. A preliminary injunction is granted early in 1972. This action would delay work on the project, but construction ultimately resumes in 1973, setting the stage for the later discovery of the snail darter and ensuing ESA litigation.

1972: The Marine Mammal Protection Act (MMPA) establishes a moratorium on the taking of all marine mammals and a ban on the importation of products derived therefrom without a permit, the issuance of which places the burden of proof on the applicant to show that their actions would not negatively affect a species' population. Today, the MMPA protects numerous species also listed under the ESA, such as whales, dolphins, seals, sea lions, sea otters, manatees, and polar bears.

1972: The Environmental Protection Agency outlaws DDT as a pesticide because of its potential danger to people. The chemical is linked to the thinning of eggshells of bald eagles and peregrine falcons, reducing hatching success and contributing to their endangered status.

1973: The Convention on International Trade in Endangered Species of Wild Fauna and Flora is signed in Washington on March 3. Eighty nations sign this treaty to protect designated plant and animal species by regulating or prohibiting international trade in certain species except by permit.

1973: The ESA of 1973 supersedes two earlier endangered species acts, broadens and strengthens protection for all plant as well as animal species listed by the United States as threatened or endangered, prohibits take and trade without a permit, requires federal agencies to avoid jeopardizing their survival, and requires actions to promote species recovery. The ESA defines an "endangered species" as any species "in danger of extinction throughout all or a significant portion of its range." A "threatened species" is one "likely to become endangered within the foreseeable future throughout all or a significant portion of its range." The ESA has become one of the most effective tools in the continuing effort to protect imperiled species and their habitats in the United States.

1973: University of Tennessee ichthyologist Dr. David Etnier discovers the snail darter, a tiny fish that appears to live only in the Little Tennessee River, which is threatened by the Tellico Dam project.

1974: What is thought to be the last wild population of black-footed ferrets (discovered only in 1964) dies out in Mellette

County, South Dakota. A small captive population dies out in 1979 due to canine distemper, causing many to fear that the species has become extinct.

1975: The Smithsonian Institution, which was directed by the ESA to identify plant species in need of ESA protection, produces a report recommending more than 3,000 plant species for possible listing as threatened or endangered. Most would never actually be listed due to limited agency resources. The small number of species actually listed compared to the species that could have been listed if unlimited resources were available illustrates a core tension of the ESA. Moreover, it is inevitable that some species will go extinct due to natural processes, and the ESA provides no means to differentiate those from species imperiled by human activities. This is an ongoing conflict that persists in the ESA to this day.

1975: Early implementing regulations for the ESA of 1973 are adopted by the Department of the Interior. They include two policy choices by the department and its attorneys that would later lead to controversy.

First, a "blanket" rule under section 4(d) of the ESA extends protections intended by Congress for endangered species to all threatened species, with limited exceptions. This reduces management flexibility for states, burdens regulated entities, and later leads to conflict around the ESA. This approach would be reversed by the Trump administration but reinstated under President Biden.

Second, section 6 of the ESA ("Cooperation with States") is narrowly interpreted and implemented as a funding mechanism only, and states are afforded no opportunity to participate in the management of listed species. This is contrary to the intent expressed by many members of Congress and Nixon administration officials, but in subsequent years, Congress and the courts alike appeared satisfied with the status quo, cementing this interpretation of the ESA into place.

Notably, Congress never took action to correct these decisions, and as a result, they became widely accepted among ESA practitioners. Controversy has instead focused on other areas of federal–state cooperation, such as when and how extensively states are consulted about ESA decisions that impact them, such as listing species or designating critical habitat.

1975: The snail darter is listed as an endangered species due to a petition filed by University of Tennessee law student Hiram Hill, his professor Zygmunt J. B. Plater, and Joseph P. Congleton of Trout Unlimited.

1975: Whooping cranes are first reintroduced to Grays Lake National Wildlife Refuge, Idaho. This population failed to reproduce in the wild, and the population died out in 2002.

1975: The grizzly bear is listed as a threatened species.

1976: Passage of the Federal Land Policy and Management Act (FLPMA) ends the federal government's century-old policy of disposing of western public lands. The new policy is to retain, protect, and manage the remaining western lands not already disposed of or incorporated into national forests, national parks, and so on; mostly low-quality rangelands remain. The Bureau of Land Management (BLM) is directed to manage these lands, with FLPMA serving as a new organic act for the agency.

FLPMA is commonly cited as the primary inspiration for the Sagebrush Rebellion, a western states'-rights movement that sought greater local control of federally managed public lands, primarily to enable resource extraction and private commercial benefits. The movement peaked in 1980 with the endorsement of presidential candidate Ronald Reagan and declined thereafter. Its influence would continue, however, inspiring states' and property rights movements, such as legal efforts to transfer federal public lands to states, opposition to federal wilderness designation, and Nevada rancher Cliven Bundy's opposition to BLM grazing regulations.

1976: Congress amends the ESA to allow commercial trade in certain articles of scrimshaw and federally owned sperm whale oil reserves. This prompt legislative response to a narrow concern is representative of the bipartisanship and productivity that characterizes Congress in the 1970s.

1976: The U.S. District Court for the Eastern District of Tennessee rules that the Tellico Dam would "adversely modify" and perhaps destroy the snail darter's critical habitat, in violation of section 7 of the ESA. However, the court allows dam construction to continue because it was already under way.

1977: Congress amends the ESA to loosen the requirements for states to access funds under section 6 of the ESA in recognition of the fact that state participation in endangered species conservation is falling short due to the 1973 law's stringent requirements. This is another example of Congress in the 1970s promptly adopting bipartisan legislation to address a problem with the ESA, but despite this attempt, ESA funding for state participation in endangered species conservation programs will remain inadequate throughout the history of the law.

1977: The first four plant species are listed as endangered: the San Clemente Island Indian paintbrush, San Clemente Island larkspur, San Clemente Island broom, and San Clemente Island bush-mallow.

1977: In *Douglas v. Seacoast Products, Inc.*, the U.S. Supreme Court characterizes state wildlife sovereignty as a legal anachronism and a nineteenth-century legal fiction but stops short of explicitly overruling the 1896 precedent *Geer v. Connecticut*.

1977: In the snail darter case, the Sixth Circuit Court of Appeals reverses the district court and orders construction of the Tellico Dam halted. This immediately attracts the attention and concern of members of Congress, especially the Tennessee delegation.

1978: In *Tennessee Valley Authority v. Hill*, the U.S. Supreme Court upholds the circuit court's decision enjoining construction of the Tellico Dam. Chief Justice Warren E. Burger writes for the Court that the "plain intent of Congress in enacting this statute [the ESA] was to halt and reverse the trend toward species extinction, *whatever the cost*. This is reflected not only in the stated policies of the Act, but in literally every section of the statute" (emphasis added).

Congress was already aware of the snail darter issue because of the Sixth Circuit's decision in 1977, but the Supreme Court decision captures the attention of the nation, with coverage in major newspapers from coast to coast. Americans are astonished that an unknown fish a few inches long can stop completion of a major public works project, and federal agencies, state governments, and industry alike are concerned that their projects, too, might be stymied by the ESA.

1978: The ESA amendments of 1978 formalize the process under section 7 of the ESA by which federal agencies consult with the FWS to ensure that their actions are not likely to jeopardize the survival of listed species or adversely modify designated "critical habitat." Further, in response to *Tennessee Valley Authority v. Hill*, Congress creates the Endangered Species Committee and empowers it to allow exemptions to this provision.

Other changes address listing, critical habitat, recovery plans, reviews of listed species every five years, and section 6 funding for plant conservation. Under listing, Congress introduces for the first time a requirement that the federal government respond to ESA petitions within ninety days, eventually imposing a significant litigation burden on the federal government.

1978: The Portland Audubon Society files an administrative challenge under the National Forest Management Act (NFMA), alleging that the U.S. Forest Service's 1977 Spotted Owl Management Plan failed to meet the NMFA's requirement that the agency maintain "viable populations of existing native and desirable non-native vertebrate species" as applied to the northern spotted owl. The Forest Service denies the administrative appeal, leading to subsequent litigation.

Ultimately, the Portland Audubon Society and the Seattle Audubon Society became the lead plaintiffs in old-growth forest management litigation against the BLM and the Forest Service, respectively, and emerge victorious after four years of litigation, six trips to the Ninth Circuit, and one decision of the U.S. Supreme Court. This forces the federal government to protect the northern spotted owl's habitat.

1979: The Endangered Species Committee exempts the Grayrocks Dam project in Wyoming from section 7 of the ESA but denies an exemption for the Tellico Dam project in Tennessee. This decision enrages the congressional delegation from Tennessee, and Representative John Duncan Sr. (R-TN) and Senator Howard Baker Jr. (R-TN) begin seeking a legislative exception specific to the Tellico Dam project.

1979: In *Hughes v. Oklahoma*, the U.S. Supreme Court explicitly reverses the landmark precedent of *Geer v. Connecticut*, with Justice Brennan writing for the Court that "the cases defining the cope of permissible state regulation in areas of congressional silence reflect an often controversial evolution of rules to accommodate federal and state interests. *Geer v. Connecticut* was decided relatively early in that evolutionary process. We hold that time has revealed the error of the early resolution reached in that case, and accordingly, *Geer* is today overruled."

1979: Over the course of the year, Representative John Duncan Sr. (R-TN) and Senator Howard Baker Jr. (R-TN) make three attempts to insert language into legislation that would force the completion of the Tellico Dam, finally succeeding on their third attempt. In September, Congress passes an appropriations bill that includes a legislative rider providing an exemption for the Tellico Dam project, flooding critical habitat of the snail darter. The fish survives, however, including in adjacent streams it had been translocated into.

1980s

1980: All surviving red wolves are taken into captivity, where a successful breeding program positions the species to be reintroduced into the wild.

1981: President Ronald Reagan's secretary of the interior, James Watt, reduces ESA listings to a crawl as part of the broader pro-industry, deregulatory approach of the administration.

1981: Black-footed ferrets are rediscovered near Meeteetse, Wyoming, ending fear that the species is extinct.

1981: The FWS adopts a policy stating that mitigating adverse impacts of federal projects on threatened and endangered species (meaning that adverse impacts are allowed to take place but are mitigated) is inconsistent with the ESA and therefore may not be used in section 7 consultation.

1982: The ESA amendments of 1982 allow, by permit, the taking of listed wildlife incidental to otherwise lawful activities provided that the permit holder implements a Habitat Conservation Plan (HCP) for the species. This provision is crafted in response to and at the request of a unique collaborative around real estate development on San Bruno Mountain outside of San Francisco.

A second major innovation is the creation of section 10(j) of the ESA, authorizing establishment of experimental populations of listed species and their management to minimize conflicts with local communities. This authority proves important for many species but also spawns controversy in the case of species such as the Mexican gray wolf and the red wolf.

The legislation also refines the ninety-day statutory deadline added to section 4 petitions in 1978 and further establishes a twelve-month deadline (from the time of a petition) for a formal listing proposal. Congress also expands the ESA's citizen suit provision to allow lawsuits to compel federal agencies to meet these deadlines. These are direct responses to the tardiness of the Reagan administration in listing species and will later birth countless listing petitions, including "megapetitions," and lawsuits. The 1982 amendments also include a prohibition against collecting listed plants on federal lands and further refinements to section 7.

1983: Reagan's interior secretary, James Watt, is forced out of office, and after his departure, ESA listings resume a more normal pace.

1983: The nation's first habitat conservation plan is approved for the protection of listed species at San Bruno Mountain, California.

1983: Creation of the Interagency Grizzly Bear Committee (IGBC), made up of bear biologists and representatives of the U.S. Forest Service, the National Park Service, the FWS, the BLM, the U.S. Geological Survey, Idaho, Montana, Washington, and Wyoming. Many controversies over bear management, depredations, landowner compensation, and delisting will arise in the future, and the IGBC will provide a collegial and effective forum for state and federal collaboration in addressing these challenges.

1984: The snail darter fish that led to the *Tennessee Valley Authority v. Hill* Supreme Court case is downlisted from endangered to threatened.

1984: In *Sierra Club v. Clark*, the U.S. Supreme Court rules that the threatened gray wolf in Minnesota could not be subject to a sport hunting season because take of a listed species can be authorized only if "population pressures within the animal's ecosystem cannot otherwise be relieved." This ruling further restricts management flexibility for species listed as threatened as opposed to endangered.

1985: The 1985 Farm Bill establishes the Conservation Reserve Program, which subsidizes the removal of erodible lands from production and the planting of cover crops to stabilize soil, improve water and air quality, and provide habitat for wildlife as well as sodbuster and swampbuster provisions as disincentives to convert natural habitat to agricultural production. It also includes a separate conservation title for the first time, which becomes a feature of every subsequent Farm Bill.

1985: The last nine remaining wild California condors are brought into captivity to prevent the species' extinction and to begin captive breeding programs at the San Diego and Los Angeles zoos.

1985: Canine distemper and plague cause the sole remaining black-footed ferret population, in Meeteetse, Wyoming, to crash. Captive breeding efforts are launched and will eventually be successful.

1986: The North American Waterfowl Management Plan, signed by the United States and Canada (Mexico joined in 1994), establishes a basis for international cooperation in bird conservation and leads to the creation of migratory bird joint ventures.

1986: A petition is filed seeking listing of the northern spotted owl under the ESA.

1987: A second such petition seeking listing of the northern spotted owl is filed.

1987: The American alligator is delisted due to recovery.

1987: All nineteen remaining black-footed ferrets are taken into captivity.

1987: The red wolf is reintroduced into the wild at Alligator River National Wildlife Refuge in eastern North Carolina.

1987: The Northern Rocky Mountain Wolf Recovery Plan calls for the translocation of gray wolves from Canada and the establishment of a new population in central Idaho and Yellowstone National Park.

1987: The FWS launches the Partners for Fish and Wildlife program.

1988: Congress makes the final major amendment to the ESA. It addresses recovery planning, mandates monitoring of delisted species for five years, increases protections for plants, and requires regular reports to Congress on progress on recovery plans and "reasonably identifiable" expenditures to conserve listed species.

1989: The oil tanker *Exxon Valdez* runs aground in Prince William Sound, Alaska, spilling 10.8 million gallons of crude oil, creating one of the largest marine oil spills in U.S. history.

1989: Ivory imports are banned in the United States to help reduce poaching of African elephants.

1989: The 1990 Interior Appropriations Act includes section 318, a legislative rider requiring the BLM and Forest Service to proceed with 4.2 billion board feet of timber sales in the next two years, notwithstanding an existing injunction issued in the northern spotted owl/old-growth forest litigation. It also establishes the Interagency Spotted Owl Committee (ISC), a panel of expert scientists charged with developing a forest management plan that would protect the northern spotted owl from extinction, led by Dr. Jack Ward Thomas.

1990s

1990: The 1990 Farm Bill establishes the Wetlands Reserve Program to repair and preserve wetlands, including those significant for wetland-dependent wildlife, which later evolves into Wetland Reserve Easements.

1990: The northern spotted owl is listed as threatened after a "not warranted" determination made in 1987 by the FWS is reversed by court order. The owl became the focal point for a bitter, acrimonious debate in the public square that pitted owls and environmental protection against human jobs and timber-dependent families and communities.

1990: The ISC issues a conservation strategy that would protect the owl by setting aside 7.7 million acres of owl habitat in large, contiguous blocks in order to avoid fragmentation, including 3 million acres that otherwise would have been subject to logging. The Bush administration largely ignores it, with BLM Director Cy Jamison proposing a crude compromise of protecting half of what the ISC recommended.

1990: The FWS proposes listing the Louisiana black bear under the ESA, prompting the Louisiana Forestry Association to convene a collaborative group of timber management companies, state and federal agencies, and NGOs to conserve the species. This "Black Bear Conservation Committee" is instrumental in the development of the FWS's recovery plan for the species as well as a section 4(d) rule eliminating conflicts between the species and normal forest management practices.

1990: The FWS opines that operations at Fort Bragg (now Fort Liberty), North Carolina, would jeopardize the continued existence of the red-cockaded woodpecker. Training operations at the base are severely curtailed, and the Department of Defense engages with the FWS and the Environmental Defense Fund, represented by Michael Bean and Robert Bonnie, to develop an innovative habitat conservation plan that provides "safe harbor" for private landowners near the base if they proactively work to establish and conserve woodpecker populations.

1991: Captive-propagated black-footed ferrets are reintroduced into the Shirley Basin of Wyoming as an experimental population several years after the last wild population was captured to prevent extinction from disease outbreaks. In the following years, ferrets will be introduced in thirty-one additional sites outside of Wyoming.

1991: California condors are reintroduced into the wild in southern California.

1991: The "Gang of Four," appointed by the House Committee on Agriculture and the House Committee on Merchant Marine and Fisheries, lays out fourteen different management options (thirty-six including sub-options) that would provide differing levels of timber harvest and protection for the northern spotted owl.

1991: The FWS issues a "jeopardy" opinion for the northern spotted owl, finding that forty-four planned BLM timber sales would jeopardize the continued existence of the species. The BLM responds by filing a petition asking the Endangered Species Committee to exempt those forty-four sales from section 7 of the ESA.

1991: Newly emboldened conservative Republicans in Congress, elected in the wake of the private property rights backlash to environmental regulation characterized by the Wise Use Movement, the County Movement, Sagebrush II, and so on, begin to introduce anti-ESA legislation with titles such as the "Human Protection Act" and the "Balanced Economic and Environmental Priorities Act."

1992: The George H. W. Bush administration imposes a moratorium on new regulations, including ESA listings. This leads to a lawsuit over the administration's failure to meet the deadline imposed in 1982, requiring petitions to be addressed within twelve months. Rather than permit the court to rule and potentially order immediate action on all 401 included species, the FWS agrees to a negotiated settlement under which it would resolve the conservation status of all 401 species by September 30, 1996, by either listing them or determining that they no longer warranted listing.

1992: The ESA's appropriations authorization lapses, and Congress fails to reauthorize the legislation. From 1992 on, the ESA functions only because annual appropriations are made each year, which are deemed to implicitly authorize continuation of the program.

1992: Civilian Department of Defense employees at Fort Benning, Georgia, are issued criminal indictments for the destruction of red-cockaded woodpecker habitat and threatened with prison sentences of up to thirty-six years and fines of up to $650,000. Charges are ultimately dropped, but the Department of Defense is further incentivized to address woodpecker conservation.

1992: The Endangered Species Committee exempts thirteen of the forty-four BLM timber sales subject to a "jeopardy" opinion in 1991, but its decision is overturned by the U.S. Supreme Court because the White House had inappropriately interfered in the committee's decision. The new Clinton administration does not renew the application for section 7 exemptions.

1993: Newly elected President Bill Clinton holds a daylong Northwest Forest Summit in Portland, Oregon, to seek a resolution to the northern spotted owl situation.

1993: The Forest Ecosystem Management Assessment Team, appointed by President Clinton, issues a report offering nine options for balancing timber harvest and northern spotted owl protection.

1993: Whooping cranes are first reintroduced to Kissimmee Prairie, Florida. This population, like the one at Gray's Lake, does not reproduce successfully and is expected to die out again.

1993: The Blackfoot Challenge is established, formalizing a decades-old landowner collaborative in Montana that engages in stream and wetland restoration, native fish restoration, access for hunters and anglers and other public recreation, noxious weed mitigation, timberland restoration, and more than 100,000 acres of conservation easements.

1994: The eastern North Pacific population of gray whales is delisted due to recovery.

1994: The Arctic peregrine falcon is delisted due to recovery.

1994: The Malpai Borderlands Group is established in southeastern Arizona and southwestern New Mexico. Subsequently, they develop a collaborative fire management plan, the first grass bank in the nation, conservation easements, and scientific best practices for range management.

1994: The FWS first offers "no surprises" guarantees for HCPs. This decision is part of a strategy devised by Secretary of the Interior Bruce Babbitt and his brain trust at the Department of the Interior to develop regulatory flexibilities to make the ESA more palatable to industry and to drive a wedge between the business community and increasingly anti-ESA Republicans in Congress.

1994: The Northwest Forest Plan is adopted, bringing the main part of the northern spotted owl/old-growth forest management controversy to a close. Subsequent litigation over the plan, individual timber sales, northern spotted owl critical habitat, and other related issues continues to the present day.

1994: The "Republican Revolution" engineered by Newt Gingrich and Dick Armey propels Republicans into power in both the House and the Senate, positioning the party to successfully oppose the Clinton administration in numerous areas.

1995: Secretary of the Interior Bruce Babbitt formalizes the Clinton administration's commitment to deconflicting the ESA with publication of the "Ten-Point Plan"—a messaging document that lays out principles for working *with* regulated industries to manage listed species.

1995: Gray wolves are reintroduced to Yellowstone National Park and central Idaho.

1995: The North Carolina sandhills red-cockaded woodpecker effort, ongoing since the early 1990s, leads to an innovative habitat conservation plan that is now recognized as the first safe harbor agreement in the nation, though safe harbor agreements wouldn't be formalized for another four years.

1995: The Carlsbad Highlands Conservation Bank, the first official agreement of its kind for a listed species, is approved for use in association with the San Diego Multiple Species Conservation Plan. In a conservation bank, landowners and financial backers together conserve wildlife habitat to generate "credits" that they can sell for a profit to those required by section 7 or 10 of the ESA to mitigate impacts on listed species. Unlike other forms of mitigation, conservation banking generates habitat *before* impacts take place, and it leverages private capital and free markets to place habitat in efficient locations.

1995: In *Sweet Home Chapter of Communities for a Great Oregon v. Babbitt*, the U.S. Supreme Court upholds the FWS regulation that defines "harm" to include destroying or modifying habitat for an endangered or threatened species if the action results in the taking of the species. This decision upholds a long-standing regulatory approach that causes landowners to fear ESA listings, making collaborative landowner conservation incentives, such as safe harbor agreements, all the more important.

1995: Congress imposes a moratorium on all ESA listings, using a legislative rider on must-pass defense appropriations legislation. The moratorium lasts for one year, putting the FWS far behind schedule on processing listing packages and vulnerable to subsequent litigation intended to enforce the listing deadlines established by Congress.

1996: The 1996 Farm Bill establishes the Wildlife Habitat Incentive Program to repair and improve fish and wildlife habitat, which later is incorporated into the Environmental Quality Incentives Program.

1996: President Bill Clinton designates the 1.7-million-acre Grand Staircase–Escalante National Monument in Utah, enraging landowners and industry in the state. Intensely controversial ever since, the monument was reduced in size by President Trump in 2017 (an event without precedent in the history of the Antiquities Act—the reduction in the size of a monument by a subsequent president) before being restored to its original size in 2021 by President Biden.

1996: The FWS and National Marine Fisheries Service finalize a joint policy to define a "distinct population segment." This policy would later play a central role in controversies over the listing status of the gray wolf.

1996: The California condor is reintroduced into northern Arizona.

1997: Bipartisan legislation to reauthorize and amend the ESA, sponsored by Republican senators Dirk Kempthorne (R-ID) and John Chafee (R-RI) and Democratic senators Max Baucus (D-MT) and Harry Reid (D-NV), is defeated by last-minute opposition from both conservative Republican members of Congress on the right (such as Richard Pombo [R-CA]) and environmental groups on the left.

1998: The first captive-bred Mexican gray wolves are released into Arizona and New Mexico.

1999: The American peregrine falcon is delisted due to recovery.

1999: The FWS formalizes both safe harbor agreements for listed species, modeled on the specialized red-cockaded woodpecker HCP developed for the North Carolina sandhills, and candidate conservation agreements with assurances, a similar mechanism for candidate species, with implementing regulations and policies. These two provisions will go on to be the premier regulatory flexibilities available to private landowners, turning threatened and endangered species habitat from a burden into something that landowners welcome and supporting conservation of dozens of species on millions of acres of land.

2000s

2000: Congress establishes State Wildlife Grants, which fund conservation of species of greatest conservation need identified in State Wildlife Action Plans. The program has since funded more than $1.128 billion in state non–game species conservation and provides the blueprint for the proposed Recovering America's Wildlife Act.

2000: The Oregon Department of Fish and Wildlife develops the nation's first candidate conservation agreement with assurances; it benefits the Columbian sharp-tailed grouse.

2001: The Aleutian Canada goose is delisted due to recovery.

2001: The Western Great Lakes Distinct Population Segment of gray wolves reaches its recovery goals but remains listed due to litigation by environmental groups seeking to keep the wolf population expanding despite its recovery.

2001: The state of Montana approaches the Blackfoot Challenge to help reduce conflicts between large predators and local livestock operators, beginning with grizzly bears.

2002: The 2002 Farm Bill establishes the Grasslands Reserve Program to restore and protect grasslands, which becomes part of the Agricultural Conservation Easement Program, and the Conservation Security Program to reward and promote conservation efforts, which becomes the Conservation Stewardship Program.

2002: The Northern Rocky Mountain Distinct Population Segment of gray wolves reaches its recovery goals but remains listed due to litigation by environmental groups seeking to keep the wolf population expanding despite its recovery.

2003: The Healthy Forests Restoration Act of 2003 establishes the Healthy Forests Reserve Program to encourage recovery actions for threatened and endangered species.

2003: Robbins' cinquefoil, a New England plant, is delisted due to recovery.

2003: The FWS adopts the Policy for Evaluation of Conservation Efforts When Making Listing Decisions, allowing consideration when listing species of conservation efforts that are planned but not yet implemented.

2003: The Department of Defense establishes the Readiness and Environmental Protection Integration Program.

2003: Congress exempts Department of Defense installations from critical habitat designation provided that the installation is operated under an approved Integrated Natural Resources Management Plan under the Sikes Act.

2004: California condors reproduce in the wild for the first time in seventeen years.

2004: The first megapetition is filed by the Center for Biological Diversity (CBD), simultaneously seeking the listing of 225 species. This megapetition (and others like it) serves to overload federal agencies and make it impossible for them to complete their work under the ESA's listing deadlines without exposing themselves to litigation. Future megapetitions from the CBD and WildEarth Guardians will include 205, 404, and 475 species.

2004: The Western Association of Fish and Wildlife Agencies begins working in earnest to conserve the greater sage-grouse. This effort results in numerous conservation initiatives and unprecedented state investment and collaboration in sage-grouse conservation.

2005: The White House Conference on Cooperative Conservation is held in St. Louis, Missouri.

2005: The House of Representatives passes Representative Richard Pombo's (R-CA) Threatened and Endangered Species Recovery Act of 2005. Defeated in the Senate, the legislation would have included economic factors in ESA decisions, abolished critical habitat, and weakened the ESA in numerous other ways. Fiercely opposed by environmentalists, the legislation contributes to Pombo's defeat in the 2006 elections. This episode is a major reason why subsequent anti-ESA legislation has served merely as political messaging to the Republican base and not been seen as serious proposals that could become law.

2007: The bald eagle is delisted following recovery, capping an eight-year process that began with President Clinton's announcement at a White House ceremony.

2007: The Greater Yellowstone Ecosystem Distinct Population Segment of grizzly bear reaches its recovery goals but remains listed due to litigation by environmental groups seeking to keep the bear population expanding, just as happened a few years earlier with the gray wolf.

2007: Deputy Assistant Secretary for Fish, Wildlife, and Parks Julie A. MacDonald is forced to resign due to allegations that she interfered with decisions made by field staff and manipulated scientific data to avoid ESA listings.

2007: Two megapetitions are filed by Forest Guardians (now WildEarth Guardians), seeking the listing of 205 and 475 species, respectively.

2007: The collaboration between the state of Montana and the Blackfoot Challenge expands to include gray wolves after they moved into the area. The group's work continues today and is an outstanding example of the effectiveness of collaboration being the key to resolving tension between conservation and private enterprise.

2008: The 2008 Farm Bill establishes the Conservation Stewardship Program, a continuation of the Conservation Security Program, and incorporates the Healthy Forests Reserve Program into the Farm Bill.

2008: The FWS, the BLM, and the Center of Excellence for Hazardous Materials Management agree to the first joint candidate conservation agreement/candidate conservation agreement with assurances; it benefits the lesser prairie chicken and the dunes sagebrush lizard.

2008: The Western Association of Fish and Wildlife Agencies enters into a memorandum of understanding with six federal agencies (the BLM, the FWS, the U.S. Forest Service, the Natural Resources Conservation Service [NRCS], the U.S. Geological Survey, and the Farm Service Agency) to facilitate sage-grouse conservation.

2008: The polar bear is listed as threatened due to habitat loss in the Arctic. Although climate change and the resulting loss of sea ice are identified as the primary threats to the species, the Bush administration concludes—and all subsequent administrations reaffirm—that the ESA is not a suitable vehicle for successfully regulating global greenhouse gas emissions.

2010s

2010: Another megapetition is filed by the CBD, seeking the listing of 404 aquatic, riparian, and wetland species in the Southeast. This petition leads directly both to the section 4 multidistrict litigation and subsequent settlements and to new collaborative conservation initiatives in the Southeast under the leadership of the regional director at the time, Cindy Dohner.

2010: Unable to handle the hundreds of species environmental groups have petitioned for listing ahead of the deadlines imposed by Congress in 1982, the FWS enters into the ESA section 4 multidistrict litigation, ultimately agreeing to make listing or critical habitat determinations for 1,030 species, subspecies, and populations by 2017. One beneficial consequence of this is the adoption of a listing workplan, providing transparency and predictability to the ESA listing program. The workplan helps spur voluntary collaborative conservation of candidate species by providing a time frame within which conservation efforts need to achieve their goals in order to avoid a species' listing.

2010: NRCS launches the Lesser Prairie Chicken Initiative and the Sage Grouse Initiative, combining Farm Bill funding for wildlife conservation with regulatory assurances under the ESA, guaranteeing that participants in the program who work to conserve listed and candidate species will not suffer any further regulations if the species are listed in the future.

2010: The Deepwater Horizon disaster in the Gulf of Mexico, the largest oil spill in history, releases 210 million gallons of oil from a damaged drilling platform and results in unprecedented multi-billion-dollar fines paid by BP for coastal restoration. Included are charges for incidental take of protected species under the Migratory Bird Treaty Act.

2011: To address the hundreds of listing determinations required by the section 4 multidistrict litigation settlements, the FWS develops a "listing workplan" for 2011–2016, laying out its agenda and providing notice of upcoming decisions to agencies, states, regulated industries, and interested parties.

2011: The Northern Rocky Mountain Distinct Population Segment of gray wolves is delisted by congressional action, except in Wyoming.

2011: Whooping cranes are first reintroduced to central Wisconsin, establishing a migratory population that winters across the southeastern United States. Also in 2011, another population is established around White Lake in Louisiana and initially appears to be successful.

2011: Governor Matthew Mead (R-WY) and Governor John Hickenlooper (D-CO) propose to Secretary of the Interior Ken Salazar a state–federal collaboration that ultimately evolves into the Sage Grouse Taskforce, a multistate collaboration with the Department of the Interior to coordinate the efforts of all eleven states in the range, all four federal land management agencies, industry, local conservation organizations, and landowners.

2011: The federal government and the fifteen southeastern states launch the Southeast Conservation Adaptation Strategy to promote conservation throughout the Southeast and the Caribbean.

2012: The FWS and NRCS launch Working Lands for Wildlife. Under this program, NRCS funds conservation of targeted species through the Farm Bill. NRCS and FWS conduct a programmatic section 7 consultation for the targeted species. Thereafter, when NRCS enrolls individual landowners into the program, they are covered by the biological opinion issued to NRCS, and the covered conservation practices are protected from any further regulation under the ESA, including both any penalties under the ESA and any obligation to carry out additional conservation.

2013: The Department of Defense establishes the Sentinel Landscapes Program identifying valuable habitat on private lands around participating military installations, and federal agencies, including the FWS, the Department of Agriculture, the Forest Service, and the BLM, prioritize those lands for funding under their conservation programs, while NGO partners work directly with landowners to implement conservation on their lands.

2013: The FWS begins reintroducing black-footed ferrets to private land using safe harbor agreements because they are more easily established than experimental populations.

2014: The 2014 Farm Bill establishes the Agricultural Conservation Easement Program, which combines the Farm and Ranch Lands Protection Program, the Grassland Reserve Program, and the Wetlands Reserve Program as well as the Regional Conservation Partnership Program, allowing partner organizations to design conservation projects using Farm Bill funding.

2014: The monarch butterfly is first petitioned for listing under the ESA.

2014: The lesser prairie chicken is listed as a threatened species.

2015: A court overturns the lesser prairie chicken's status as threatened.

2015: The greater sage-grouse is determined not to warrant listing due to the largest collaborative conservation effort in American history, an investment that will eventually top $1.5 *billion*. Federal, state, and private organizations were all involved in myriad conservation programs, with over half of the species' range secured by BLM Resource Management Plan amendments.

2015: The Obama administration greatly enlarges the Mexican gray wolf recovery area in Arizona and New Mexico, sparking tensions with state governments and local residents who object to their proximity.

2015: The National Alliance of Forest Owners launches its Wildlife Conservation Initiative, a program under which its members manage their 46 million acres of private land to produce and maintain young forest, open-canopy forests, and healthy riparian and aquatic habitats to benefit wildlife, even while managing their lands as working forests, to achieve both economic and environmental ownership objectives.

2016: The Louisiana black bear is delisted, in large part thanks to the work done by the Black Bear Conservation Committee, which is a model for future collaborative conservation efforts.

2016: President Obama designates the 1.3-million-acre Bears Ears National Monument in Utah. Like Clinton's Grand Staircase–Escalante National Monument designation in 1996, this is widely supported by environmentalists across the nation but angers the state of Utah and large numbers of westerners opposed to the expansion of protections on federal lands in the West.

2016: The Obama administration updates the 1981 mitigation policy, recognizing that it has been inconsistent with the ESA since the 1982 amendments allowing mitigation under habitat conservation plans. Since 1982, the FWS has used mitigation in not only habitat conservation plans but also other section 10 permits and section 7 consultations, effectively ignoring the 1981 policy.

2016: The FWS and National Marine Fisheries Service issue new regulations restricting ESA-listing petitions to a single species and imposing clear requirements for the quality of the information presented in the petition. This serves the dual purposes of eliminating the megapetitions for hundreds of species that inundated the agencies in the 2000s and providing the agencies with better information for their decisions.

2016: The FWS formalizes "species status assessments" as a tool for collecting the best available scientific information for both listing decisions and subsequent five-year status reviews.

2016: Captive-bred black-footed ferrets are reintroduced to Meeteetse, Wyoming, where the species had been rediscovered in 1981.

2016: Legendary ecologist Edward O. Wilson publishes *Half-Earth*, calling on the world to protect half of Earth in order to preserve 85 to 90 percent of species and maintain the biodiversity that all life, including human, depends on.

2016: The Recovering America's Wildlife Act is first introduced in the House of Representatives. Reintroduced in every Congress since, the legislation passes the House in both 2020 and 2022 but has yet to pass the Senate. In its current form, it would provide almost $1.4 billion per year for state and tribal governments to conserve nongame, threatened, endangered, and at-risk species.

2017: The gray wolf is delisted in Wyoming.

2017: The rusty-patched bumblebee is listed as an endangered species, the first bee in the forty-eight contiguous states to be listed.

2017: President Donald Trump reduces the size of Utah's Grand Staircase–Escalante National Monument by half and Utah's Bears Ears National Monument by 85 percent. This is the first time a president has used the Antiquities Act to *shrink* an existing national monument, and the decision is met by skepticism and criticism from environmentalists but praise from many western state lawmakers.

2017: The Trump administration issues a legal opinion stating that the Migratory Bird Treaty Act does not include criminal liability for incidental take, contrary to past enforcement actions, such as the fines levied against BP following the 2010 Deepwater Horizon disaster.

2017: The Trump administration revokes the Obama-era withdrawal from mineral leasing of all sagebrush focal areas.

2017: Conservation Without Conflict is launched as a formal coalition to build on the success of voluntary collaborative conservation in the Southeast and Northeast. The organization embraces the importance of private land for fish and wildlife conservation and the necessity of working collaboratively with landowners to achieve conservation. It embraces incentives for landowners, the best available science, economic and cultural benefits of working landscapes, outdoor recreation including hunting and fishing, and early, proactive conservation interventions.

2018: The 2018 Farm Bill maintains existing programs, providing $6 billion each year for conservation on private working lands. Although wildlife is only one of the resources prioritized for funding under the various Farm Bill programs, the law nonetheless is the single largest source of funding for wildlife conservation in America.

2018: The Trump administration issues amended BLM Resource Management Plans to reduce the regulatory burden of greater sage-grouse conservation on states and industry.

2018: The Trump administration rescinds the 2016 Obama policy on mitigation, seemingly for no reason other than it having been adopted by Obama. This restores the outdated 1981 policy.

2018: In *Weyerhaeuser v. U.S. Fish and Wildlife Service*, the U.S. Supreme Court holds that critical habitat cannot be designated in an area unless that area is in fact "habitat" for the species, which needs to be defined.

2019: A paper published in the journal *Science* announces that North America has lost almost 3 *billion* birds since 1970, sparking growing awareness of the worldwide biodiversity crisis and its impact on birds, insects, forests, and other components of Earth's biosphere.

2019: The FWS issues a major package of ESA regulations, abolishing the blanket 4(d) rule that dated back to 1975, limiting designation of critical habitat in unoccupied areas, and streamlining section 7 consultations. Although many of these changes will later be reversed by the Biden administration, most of the section 7 changes will endure because they provide additional transparency and predictability for regulated entities and/or faster permitting without exposing endangered species to any harms.

2019: The FWS's whooping crane program at Patuxent, Maryland, is shut down after fifty-five years. Captive breeding continues at other facilities.

2020s

2020: The FWS adopts its first *delisting* workplan, providing the same public notice for delisting determinations as for listing determinations.

2020: The Trump administration issues a regulatory definition of "habitat" consistent with the Supreme Court's opinion in *Weyerhaeuser v. U.S. Fish and Wildlife Service*.

2020: The gray wolf is delisted nationwide.

2020: A second petition is filed for listing of the monarch butterfly. The butterfly is found to warrant listing due to a decline of 90 percent in its western population and 90 percent in its eastern population, but the listing is precluded by higher-priority actions to amend the Lists of Endangered and Threatened Wildlife and Plants.

2021: The Trump administration issues regulations implementing its position that the Migratory Bird Treaty Act does not include criminal liability for incidental take.

2021: President Joe Biden restores Grand Staircase–Escalante National Monument and Bears Ears National Monument in Utah to their original sizes.

2021: The Biden administration rescinds both the legal opinion and the regulations issued by the Trump administration providing that the Migratory Bird Treaty Act does not include criminal liability for incidental take. This action restores the historic practice of seeking criminal penalties for the incidental take of migratory birds.

2021: The Biden administration begins the process of once again revising the BLM Resource Management Plans designed to conserve the greater sage-grouse.

2021: A federal district court overturns the Trump administration's decision to revoke the Obama-era withdrawal from mineral leasing of all sagebrush focal areas.

2021: The states of Idaho, Montana, and Wisconsin all adopt aggressive gray wolf hunting policies with the intent of reducing their wolf populations. As a result of these controversial policies, the FWS initiates a status review of the northern Rocky Mountain population of the wolf, which is ongoing.

2021: The Western Association of Fish and Wildlife Agencies issues a report warning that greater sage-grouse populations continue to decline. A follow-up report outlining next steps is expected in the near future.

2022: A federal district court overturns the nationwide delisting of the gray wolf, leaving only wolves in Idaho, Montana, and Wyoming delisted.

2022: The snail darter, made famous by the Tellico Dam and *Tennessee Valley Authority v. Hill*, is delisted due to recovery and a petition filed by Zygmunt J. B. Plater, following nine years as an endangered species and thirty-eight years as a threatened species.

2023: The Biden administration rescinds the Trump administration's regulatory definition of "habitat."

2023: The Biden administration adopts a revised form of the 2016 Obama mitigation policy, suggesting a troubling pattern of tit-for-tat regulatory actions, where the political party in the White House may be becoming more important than either the language of the ESA (the law) or what's best for species.

2023: The lesser prairie chicken is again listed, this time as threatened in one part of its range and endangered in another. Congress attempts to invalidate the listing through use of the Congressional Review Act, but President Biden vetoes the legislation.

Appendix B

Federal Environmental and Consumer Protection Statutes and Agencies Established during the 1960s and 1970s Green Revolution

Year	Environmental Legislation	Consumer Protection Legislation
1963	Clean Air Act (amended 1965, 1966, 1969, 1970, 1977, 1990) Federal Water Quality Control Act	
1964	Wilderness Act Land and Water Conservation Fund Act	
1965	Water Quality Act of 1965 National Emissions Standards Act Motor Vehicle Air Pollution Control Act Solid Waste Disposal Act	
1966	Fur Seal Act Clean Water Restoration Act Endangered Species Act (amended 1969, 1973)	Motor Vehicle Safety Act Federal Cigarette Labeling and Advertising Act Child Protection Act of 1966 Fair Packaging and Labeling Act of 1966
1967	Air Quality Act	Flammable Fabrics Act of 1967 Wholesome Poultry Products Act of 1967 Wholesome Meat Act of 1967
1968	National Trails System Act Wild and Scenic Rivers Act	Federal Consumer Credit Protection Act Natural Gas Pipeline Safety Act of 1968 Radiation Control Act of 1968

1969

Child Protection and Toy
   Safety Act
Coal Mine Health and Safety Act
   of 1969
Fire Research and Safety Act
   of 1969

1970   National Environmental
     Policy Act
   Environmental Quality
     Improvement Act
   Environmental Protection Agency
     created
   Council on Environmental Quality
     created

Fair Credit Reporting Act of 1970
Lead-Based Paint Poisoning
   Prevention Act
Occupational Safety and Health
   Administration agency (OSHA)
   created

1971   Wild Free-Roaming Horses and
     Burros Act

1972   Marine Mammal Protection Act
   Coastal Zone Management Act
   Federal Water Pollution Control
     Act Amendments of 1972
     (Clean Water Act)
   Marine Protection, Research
     and Sanctuaries Act of 1972
     (Ocean Dumping Act)

Federal Insecticide, Fungicide,
   and Rodenticide Act
Consumer Products Safety
   Commission created
Noise Control Act

1973   Endangered Species Act
     (amended 1978, 1982, 1988)

1974   National Reserves
     Management Act
   Safe Drinking Water Act
   Forest and Rangeland Renewable
     Resources Planning Act
     of 1974

Emergency Highway Energy
   Conservation Act

1975   Energy Policy and
     Conservation Act

Magnuson-Moss Warranty Act
Hazardous Materials
   Transportation Act

1976   National Forest Management Act
   Federal Land Policy and
     Management Act
   Resource Conservation and
     Recovery Act
   Toxic Substances Control Act
   Magnuson-Stevens Fisheries
     Conservation and
     Management Act

1977   Surface Mining Control and
            Reclamation Act
        Department of Energy created
        U.S. Strategic Petroleum Reserve
            established

1978   National Energy Conservation
                                                    Policy Act
                                                Power Plant and Industrial Fuel
                                                    Use Act of 1978
                                                Natural Gas Policy Act of 1978
                                                Public Utilities Regulatory
                                                    Policies Act of 1978
                                                Energy Tax Act of 1978

1980   Equal Access to Justice Act
        Comprehensive Environmental
            Response, Compensation, and
            Liability Act
        Alaska National Interest Lands
            Conservation Act
        Fish and Wildlife
            Conservation Act

Notes

Introduction

1. NatureServe, "Over One-Third of Biodiversity in the United States Is at Risk of Disappearing," press release, February 6, 2023, https://www.natureserve.org/news-releases/over-one-third-biodiversity-united-states-risk-disappearing.

2. Kenneth V. Rosenberg et al., "Decline of the North American Avifauna," *Science* 366, no. 6461 (2019), https://doi.org/DOI: 10.1126/science.aaw1313.

3. "Dramatic Decline in Western Butterfly Populations Linked to Fall Warming," *ScienceDaily*, March 4, 2021, https://www.sciencedaily.com/releases/2021/03/210304145405.htm; M. L. Forister et al., "Fewer Butterflies Seen by Community Scientists across the Warming and Drying Landscapes of the American West," *Science* 371, no. 6533 (March 5, 2021), https://doi.org/10.1126/science.abe5585; Xerces Society, "Monarchs in Decline," accessed September 18, 2023, https://xerces.org/monarchs/conservation-efforts; Liz Kimbrough, "Western Monarch Populations Reach Highest Number in Decades," *Mongabay*, January 31, 2023, https://news.mongabay.com/2023/01/monarch-populations-rebound-but-its-still-a-long-journey-to-recovery.

4. Nathalie Steinhauer, "United States Honey Bee Colony Losses 2022–23: Preliminary Results from the Bee Informed Partnership," *Bee Informed*, June 22, 2023, https://beeinformed.org/2023/06/22/united-states-honey-bee-colony-losses-2022-23-preliminary-results-from-the-bee-informed-partnership. These commercial honeybees are bred and then moved around the country to pollinate crops such as almonds, apples, blueberries, cherries, pumpkins, and many others, including fruits, nuts, vegetables, legumes, oilseeds, and forage crops.

5. Kelsey Kopec and Lori Ann Burd, *Pollinators in Peril, Center for Biological Diversity*, February 2017, 1, https://www.biologicaldiversity.org/campaigns/native_pollinators/pdfs/Pollinators_in_Peril.pdf.

6. Edward O. Wilson, *The Diversity of Life* (Cambridge, MA: Harvard University Press, 1992), 287–88; Healy Hamilton, "The Role of Biodiversity in a Resilient America," presentation at Wildlife Corridors and Saving America's Biodiversity with Edward O. Wilson (Washington, DC), October 24, 2017.

7. Hamilton, "The Role of Biodiversity in a Resilient America."

8. Wilson, *The Diversity of Life*, 283–87; Carrie Arnold, "Horseshoe Crab Blood Is Key to Making a COVID-19 Vaccine—But the Ecosystem May Suffer," *National Geographic*,

July 2, 2020, https://www.nationalgeographic.com/animals/article/covid-vaccine-needs -horseshoe-crab-blood; *Alabama–Tombigbee River Coalition v. Kempthorne*, 477 F.3d 1250 (11th Cir. 2007); *National Association of Home Builders v. Babbitt*, 130 F.3d 1041 (D.C. Cir. 1997), 1052–53.

9. Angela Colbert, "A Global Biodiversity Crisis: How NASA Satellites Help Track Changes to Life on Earth," *National Aeronautics and Space Administration*, May 22, 2023, https://climate.nasa.gov/news/3265/a-global-biodiversity-crisis-how-nasa-satellites -help-track-changes-to-life-on-earth.

10. Environmental Protection Agency, "Future of Climate Change," January 19, 2017, https://19january2017snapshot.epa.gov/climate-change-science/future-climate-change_ .html; National Ocean Service, "2022 Sea Level Rise Technical Report," accessed July 20, 2023, https://oceanservice.noaa.gov/hazards/sealevelrise/sealevelrise-tech-report.html.

11. *National Security Strategy*, October 12, 2022, 27, https://www.whitehouse.gov/wp -content/uploads/2022/10/Biden-Harris-Administrations-National-Security-Strategy -10.2022.pdf.

CHAPTER 1

1. Noah Greenwald, Kieran F. Suckling, Brett Hartl, and Loyal A. Mehrhoff, "Extinction and the U.S. Endangered Species Act," *PeerJ* 7 (2019): e6803, https://pubmed.ncbi .nlm.nih.gov/31065461.

2. Forest2Market, "New Report Details the Economic Impact of US Forest Products Industry," May 8, 2019, https://www.forest2market.com/blog/new-report-details-the -economic-impact-of-us-forest-products-industry. This report was commissioned by the National Alliance of Forest Owners, a critical partner in many collaborative conservation programs described in this book.

3. Forest2Market, "New Report Details the Economic Impact of US Forest Products Industry"; for example, South Carolina Forestry Commission, "Forest Management Facts," accessed November 16, 2021, https://www.state.sc.us/forest/refmgt.htm.

4. Alyson C. Flournoy, "Beyond the Spotted Owl Problem: Learning from the Old-Growth Controversy," *Harvard Environmental Law Review* 17 (1993): 301.

5. 90-Day Petition Finding and Initiation of Status Review, Northern Spotted Owl, 52 Fed. Reg. 34396 (September 11, 1987).

6. Finding on Northern Spotted Owl Petition, 52 Fed. Reg. 48552 (December 23, 1987).

7. *Northern Spotted Owl v. Hodel*, 716 F.Supp. 479 (W.D. Wash. 1988).

8. Determination of Threatened Status for the Northern Spotted Owl, 55 Fed. Reg. 26114 (June 26, 1990).

9. Determination of Threatened Status for the Northern Spotted Owl, 55 Fed. Reg. 26114, 28119.

10. Eric Loomis, "Think the Spotted Owl Is to Blame for Job Losses? Think Again," *Washington Post*, September 13, 2019, https://www.washingtonpost.com/outlook/2019 /09/13/think-spotted-owl-is-blame-job-losses-think-again.

11. Shannon Petersen, "The Modern Ark: A History of the Endangered Species Act" (PhD diss., University of Wisconsin, Madison, 2000), 183–94; James Morton Turner and

Andrew C. Isenberg, *The Republican Reversal: Conservatives and the Environment from Nixon to Trump* (Cambridge, MA: Harvard University Press, 2018), 85–86.

12. Jack Ward Thomas et al., *A Conservation Strategy for the Northern Spotted Owl: Report of the Interagency Scientific Committee to Address the Conservation of the Northern Spotted Owl* (1990); Flournoy, "Beyond the Spotted Owl Problem," 295–96.

13. Flournoy, "Beyond the Spotted Owl Problem," 296–97.

14. Petersen, "The Modern Ark," 203; H. Dale Hall, *Compelled: From the Yazoo Pumps to Polar Bears and Back: The Evolution of Natural Resource Conservation and Law* (Memphis, TN: Ducks Unlimited, Inc., 2022), 177–79.

15. K. Norman Johnson et al., *Alternatives for Management of Late-Successional Forests of the Pacific Northwest: A Report to the Agricultural Committee and the Merchant Marine Committee of the U.S. House of Representatives* (October 8, 1991), 31–32.

16. Donald J. Barry, interview by Christopher E. Segal, November 2, 2021.

17. Flournoy, "Beyond the Spotted Owl Problem," 296–97.

18. Bruce G. Marcot and Jack Ward Thomas, *Of Spotted Owls, Old Growth, and New Policies: A History since the Interagency Scientific Committee Report*, General Technical Report PNW-GTR-408 (Portland, OR: U.S. Department of Agriculture, Forest Service, Pacific Northwest Research Station, 1997), 6.

19. Tom Kenworthy, "Spotted Owl in the Hands of 'God Squad,'" *Washington Post*, November 20, 1991, https://www.washingtonpost.com/archive/politics/1991/11/20/spotted-owl-in-the-hands-of-god-squad/097ea868-67b2-44b3-a554-5a17464e1b46.

20. Hall, *Compelled*, 166–86; Flournoy, "Beyond the Spotted Owl Problem," 297–98; Marcot and Thomas, *Of Spotted Owls, Old Growth, and New Policies*, 6.

21. Timothy J. McNulty and Carol Jouzaitis, "Bush, Clinton Try to Balance the Environment and Economy," *Chicago Tribune*, September 15, 1992, https://www.chicagotribune.com/news/ct-xpm-1992-09-15-9203240170-story.html; Petersen, "The Modern Ark," 221; *Timber Wars: Ep. 5: The Plan*, narrated by Aaron Scott, *Oregon Public Broadcasting*, podcast audio, 38:09, https://www.npr.org/podcasts/906829608/timber-wars.

22. Flournoy, "Beyond the Spotted Owl Problem," 309, 311n311; Turner and Isenberg, *The Republican Reversal*, 86.

23. Turner and Isenberg, *The Republican Reversal*, 86.

24. Hall, *Compelled*, 210–22; Marcot and Thomas, *Of Spotted Owls, Old Growth, and New Policies*, 10–12; Byron W. Daynes and Glen Sussman, *White House Politics and the Environment: Franklin D. Roosevelt to George W. Bush* (College Station: Texas A&M University Press, 2010), 102–3; Peter Honey, "Clinton's 'Timber Summit' to Draw Huge Crowds," *Baltimore Sun*, April 2, 1993, https://www.baltimoresun.com/news/bs-xpm-1993-04-02-1993092024-story.html; Hugh Dellios, "Concern at Timber Summit," *Chicago Tribune*, April 2, 1993, https://www.chicagotribune.com/news/ct-xpm-1993-04-02-9304020374-story.html.

25. Marcot and Thomas, *Of Spotted Owls, Old Growth, and New Policies*, 11–12; Lita P. Buttolph et al., *Northwest Forest Plan—The First 10 Years (1994–2003): Socioeconomic Monitoring of the Olympic National Forest and Three Local Communities*, General Technical Report PNW-GTR-679 (U.S. Forest Service, July 2006), iii, https://www.fs.fed.us/pnw

/pubs/pnw_gtr679.pdf; Hall, *Compelled*, 225–27; U.S. Forest Service, "Northwest Forest Plan Overview," accessed April 19, 2021, https://www.fs.fed.us/r6/reo/overview.php; U.S. Forest Service, "Northwest Forest Plan Aquatic Conservation Strategy," accessed June 3, 2021, https://www.fs.fed.us/r6/reo/acs; Turner and Isenberg, *The Republican Reversal*, 86–87.

26. Otis L. Graham Jr., *Presidents and the American Environment* (Lawrence: University Press of Kansas, 2015), 313; Turner and Isenberg, *The Republican Reversal*, 79; Susan Elderkin, "What a Difference a Year Makes," *High Country News*, September 2, 1996, https://www.hcn.org/issues/89/2748; Trilby C. E. Dorn, "Logging without Laws: The 1995 Salvage Logging Rider Radically Changes Policy and the Rule of Law in the Forests," *Tulane Environmental Law Journal* 9, no. 2 (1996): 472.

27. Michael Doyle, "Study Brings Owl-Related Job Losses Down to Earth," *Greenwire*, June 30, 2021, https://www.eenews.net/greenwire/stories/1063736201/print; Lowell E. Baier, *Inside the Equal Access to Justice Act: Environmental Litigation and the Crippling Battle over America's Lands, Endangered Species, and Critical Habitats* (Lanham, MD: Rowman & Littlefield, 2016), 352–53.

28. Determination of Critical Habitat for the Northern Spotted Owl, 57 Fed. Reg. 1796 (January 15, 1992), 1809.

29. For example, Susan Jane Brown, "The Return of the Spotted-Owl Wars?" *Seattle Times*, January 22, 2021, https://www.seattletimes.com/opinion/the-return-of-the-spotted-owl-wars ("Career agency biologists have for years warned that the spotted owl—despite herculean efforts—continues to slide toward extinction. Fewer and fewer spotted owls are reproducing at sufficient rates to sustain the species, which is functionally extinct in the northern part of its range in Canada").

30. Kenneth M. Murchison, *The Snail Darter Case: TVA versus the Endangered Species Act* (Lawrence: University Press of Kansas, 2007), 14–15, 18; *Dialogue*, "Revisiting the Tennessee Story of the Snail Darter," hosted by Chrissy Keuper, featuring Wayne Starnes, Zygmunt Plater, Peggy Shute, J. R. Shute, and Charles Sims, October 6, 2021, *WUOT*, radio program, 51:04, https://www.wuot.org/news/2021-10-06/dialogue-revisiting-the-tennessee-story-of-the-snail-darter.

31. *Dialogue*, "Revisiting the Tennessee Story of the Snail Darter"; Tommy Millsaps, "A Look Back: Closing the Tellico Dam Gates," *Monroe County Advocate & Democrat* (Sweetwater, TN), November 30, 2009, https://www.advocateanddemocrat.com/news/article_1d20abdc-a6e6-5006-9931-389bbe40538e.html; Wikipedia, "Chota (Cherokee Town)," accessed May 18, 2023, https://en.wikipedia.org/wiki/Chota_(Cherokee_town).

32. *Environmental Defense Fund v. Tennessee Valley Authority*, 339 F. Supp. 806 (E.D. Tenn. 1972); *Environmental Defense Fund v. Tennessee Valley Authority*, 371 F. Supp. 1004 (E.D. Tenn. 1973).

33. *Dialogue*, "Revisiting the Tennessee Story of the Snail Darter."

34. *Hill v. Tennessee Valley Authority*, 419 F. Supp. 753 (E.D. Tenn 1976).

35. *Hill v. Tennessee Valley Authority*, 549 F.2d 1064 (6th Cir. 1977); *Dialogue*, "Revisiting the Tennessee Story of the Snail Darter."

36. *Tennessee Valley Authority v. Hill*, 437 U.S. 153 (1978).

37. Endangered Species Act Amendments of 1978, Pub. L. No. 95–632, 92 Stat. 3751 (1978).

38. Murchison, *The Snail Darter Case*, 155–65.

39. Energy and Water Development Appropriation Act, 1980, Pub. L. No. 96–69, 93 Stat. 437, 449–50 (1979).

40. Margot Hornblower, "Carter Signs Bill Forcing Tellico Dam Completion," *Washington Post*, September 26, 1979, https://www.washingtonpost.com/archive/politics/1979/09/26/carter-signs-bill-forcing-tellico-dam-completion/7e57e3c0-d186-4bcf-9930-842c07e21c81.

41. U. S. Fish and Wildlife Service, *Snail Darter Recovery Plan, Second Revision* (December 12, 1982), 14–16.

42. Murchison, *The Snail Darter Case*, 168, 174–75.

43. Murchison, *The Snail Darter Case*, 184; Center for Biological Diversity, James D. Williams, and Zygmunt Plater, Before the Secretary of the Interior: Petition to Delist the Snail Darter under the Endangered Species Act (July 16, 2019).

CHAPTER 2

1. Alston Chase, *In a Dark Wood: The Fight over Forests and the Rising Tyranny of Ecology* (New York: Houghton Mifflin Company, 1995), 93–94. For an eloquent and beautifully written work that places the 1973 ESA within the context of the ideological development of man's thought process on nature and man's relationship thereto going back to Plato and up through the great thinkers and legendary scientists and philosophers of the eighteenth through twentieth centuries, read Dr. Alston Chase's book (pp. 94–130). Barry Commoner's cybernetic vision provides a similar result in his book *The Closing Circle: Nature, Man, and Technology* (New York: Alfred A. Knopf, 1971); see also Steven Lewis Yaffee, *Prohibitive Policy: Implementing the Federal Endangered Species Act* (Cambridge, MA: MIT Press, 1982), 37–38; Shannon Petersen, "Congress and Charismatic Megafauna: A Legislative History of the Endangered Species Act," *Environmental Law* 29 (1999): 478–79; Charles C. Mann and Mark L. Plummer, *Noah's Choice: The Future of Endangered Species* (New York: Alfred A. Knopf, 1995), 160–62; 119 Cong. Rec. 19,138 (1973) (statement of Harrison Williams); 119 Cong. Rec. 30,166 (1973) (statement of Frank Annunzio); and 119 Cong. Rec. 42,912–13 (1973) (statement of Ted Stevens).

2. Robert L. Fischman, Vicky J. Meretsky, and Matthew P. Castelli, "Collaborative Governance under the Endangered Species Act: An Empirical Analysis of Protective Regulations," *Yale Journal on Regulation* 38 (2021): 978.

3. Douglas Brinkley, *Silent Spring Revolution: John F. Kennedy, Rachel Carson, Lyndon Johnson, Richard Nixon, and the Great Environmental Awakening* (New York: Harper, 2022).

4. Samuel P. Hays, *A History of Environmental Politics since 1945* (Pittsburgh, PA: University of Pittsburgh Press, 2000), 5, 137–53.

5. Rachel Carson, *Silent Spring* (Boston: Houghton Mifflin Company, 1962). For the underlying themes of societal unrest throughout the twentieth century leading up to this era of environmental awareness, read Chad Montrie, *The Myth of Silent Spring: Rethinking the Origins of American Environmentalism* (Berkeley: University of California Press, 2018),

and Montrie's earlier book *A People's History of Environmentalism in the United States* (New York: Continuum Press, 2011).

6. National Wildlife Federation, "Conservation News," June 1, 1969.

7. Meir Rinde, "Richard Nixon and the Rise of American Environmentalism," *Distillations* (blog), *Science History Institute*, June 2, 2017, https://www.sciencehistory .org/distillations/richard-nixon-and-the-rise-of-american-environmentalism (quoting J. Brooks Flippen).

8. Hays, *A History of Environmental Politics since 1945*, 5, 22–26, 94–108; James T. Patterson, *Restless Giant: The United States from Watergate to Bush v. Gore* (Oxford: Oxford University Press, 2005), 117. For the development of the representative nongovernmental organizations (NGOs), see Baier, *Inside the Equal Access to Justice Act*, 135–270; see also James A. Tober, *Wildlife and the Public Interest: Nonprofit Organizations and Federal Wildlife Policy* (New York: Praeger Publishers, 1989), 1–58, 159–97.

9. John D. Dingell Jr., interview by Lowell E. Baier, April, 29, 2016; Judith A. Layzer, *Open for Business: Conservatives' Opposition to Environmental Regulation* (Cambridge, MA: MIT Press, 2012), 31–82.

10. Nathaniel Reed's opinion is that Nixon's motivation was building a legacy and a "really fine environmental record" rather than a voting issue to pick up the "green" vote. Nathaniel P. Reed, interview by Lowell E. Baier and Christopher E. Segal, May 18, 2016. See also John C. Whitaker, *Striking a Balance: Environment and Natural Resources Policy in the Nixon-Ford Years* (Washington, DC: AEI-Hoover Policy Studies, 1976), 1, 6; Rick Pearlstein, *Nixonland* (New York: Scribner, 2008), 460–61; J. Brooks Flippen, *Nixon and the Environment* (Albuquerque: University of New Mexico Press, 2000), 9–14, 48–49, 83–84, 129, 161–63, 177–84; John Ehrlichman, *Witness to Power: The Nixon Years* (New York: Simon and Schuster, 1982), 208; Russell E. Train, *Politics, Pollution and Pandas* (Washington, DC: Island Press, 2003), 110–11; Elizabeth Drew, *Richard M. Nixon* (New York: Time Books/Henry Holt and Company, 2007), 52–53; Lee M. Talbot, interview by Lowell E. Baier and Christopher E. Segal, March 29, 2016; and James Rathlesberger, ed., *Nixon and the Environment: The Politics of Devastation* (New York: The Village Voice, 1972), 257–69.

11. Pearlstein, *Nixonland*, 460–61; Flippen, *Nixon and the Environment*, 16; Mann and Plummer, *Noah's Choice*, 156.

12. Mary Graham, *The Morning after Earth Day* (Washington, DC: Brookings Institution, 1999).

13. Rinde, "Richard Nixon and the Rise of American Environmentalism."

14. Flippen, *Nixon and the Environment*, 9–14, 35–36, 47–49, 83–84, 129, 160–61, 177–84; Mann and Plummer, *Noah's Choice*, 157. Nixon also tasked CEQ Chairman Russell Train with developing a conservation agenda he could pursue. For a short, concise, and well-documented account of Nixon and the environmental movement, see Rinde, "Richard Nixon and the Rise of American Environmentalism."

15. Chase, *In a Dark Wood*, 84–89.

16. Talbot, interview; E. U. Curtis "Buff" Bohlen, interview by Lowell E. Baier and Christopher E. Segal, August 7, 2019, and December 12, 2020; Douglas P. Wheeler, interview by Lowell E. Baier and Christopher E. Segal, March 24, 2017, and November

23, 2021; Nathaniel Pryor Reed, *Travels on the Green Highway: An Environmentalist's Journey* (Hobe Sound: Reed Publishing Company, 2016), 215; Endangered Species Coalition, "Lee M. Talbot, a Personal Perspective on the Endangered Species Act of 1973 (ESA) and the Convention on International Trade in Endangered Species of Wild Fauna and Flora (CITIES)," accessed July 28, 2018, http://www.endangered.org/campaigns/wild-success-endangered-species-act-at-40/lee-talbot; "Address by Lee M. Talbot, Senior Scientist, Council on Environmental Quality, upon Receipt of the Schweitzer Centenary Medal," *Animal Welfare Institute Information Report* 24, no. 1 (January–March 1975); Talbot, interview.

17. Reed, interview.

18. Bohlen, interview; Talbot, interview; Reed, interview.

19. Mann and Plummer, *Noah's Choice*, 156; Wheeler, interview; Bohlen, interview.

20. Chase, *In a Dark Wood*, 81.

21. Talbot, interview.

22. Talbot, interview.

23. Frank Potter, interview by Lowell E. Baier, November 20 and 23, 2021.

24. Reed, *Travels on the Green Highway*, 214. See also the U.S. Supreme Court's later 1978 decision on this legislative language history comparing the 1966 and 1969 acts with the 1973 act. *Tennessee Valley Authority v. Hill*, 437 U.S. 173, 181–83.

25. For a very good rendition of the ESA's progress through Congress, read Dr. Alston Chase's excellently researched book *In a Dark Wood*, 79–94; Mann and Plummer, *Noah's Choice*, 157–58. The *Noah's Choice* reference is used with some degree of reservation given that the book's authors have a very biased view of environmentalists.

26. Reed, interview; Bohlen, interview; Wheeler, interview.

27. Dingell, interview; Talbot, interview, plus an unpublished document provided by Talbot titled "Patience and Tenacity: The Endangered Species Act of 1973," recalling many of the intimate and inside details of the drafting of the ESA; Bohlen, interview; Reed, *Travels on the Green Highway*, 210–15; Mann and Plummer, *Noah's Choice*, 154–61. For a persuasive argument that despite this, in fact, the FWS over the following years was able to and frequently did use its discretion to weaken the ESA; see Oliver A. Houck, "The Endangered Species Act and Its Implementation by the U.S. Departments of Interior and Commerce," *University of Colorado Law Review* 64 (1993): 298–99.

28. Just in the past decade, the science and technology protecting, recovering, and managing at-risk and endangered species has so advanced that the language of the 1973 act and its successive amendments has constrained the administrative process notwithstanding enlightened regulations promulgated to implement contemporary recovery techniques. A challenge amid the partisan politics of Washington in the twenty-first century is to get surgically crafted technical amendments made to the 1973 act by Congress so that it can reflect and be adapted to today's science and technology and environmental changes, such as the evolving consequences of climate change. Alternatively, it's up to the FWS to promulgate regulations to adapt the ESA to current science and technology.

29. Mann and Plummer, *Noah's Choice*, 157.

30. 119 Cong. Rec. 30,174 (1973); Endangered Species Act of 1973, Pub. L. No. 93–205, 87 Stat. 884 (1973); *Committee on Conference, Endangered Species Act of 1973*, Endangered Species Act of 1973, H.R. Rep. No. 93–740 (1973) (Conf. Rep.).

31. 119 Cong. Rec. 42,529, 42,534 (1973).

32. 119 Cong. Rec. 42,910–16 (1973).

33. Reed, *Travels on the Green Highway*, 215. The media gave little attention to the congressional enactment or the president signing the bill into law. The *New York Times*, the *Washington Post*, and the *Los Angeles Times* each covered it with one sentence sandwiched into an unrelated article, and the *Chicago Tribune* ignored it. "President Signs Manpower Bill," *New York Times*, December 29, 1973, 13; Austin Scott, "Nixon Signs Bill to Give States Manpower Funds," *Washington Post*, December 29, 1973, A1; "President Signs Bill Reshaping Federal Manpower Programs," *Los Angeles Times*, December 29, 1973, 1; Lewis G. Regenstein, "A History of the Endangered Species Act of 1973 and an Analysis of Its History; Strengths and Weaknesses; Administration; and Probable Future Effectiveness" (MA thesis, Emory University, 1975), 143–45.

CHAPTER 3

1. Stuart Rothenberg, "Utah's Frank Moss Was a Symbol of Nation's Realignment," *Roll Call*, February 5, 2003, https://rollcall.com/2003/02/05/utahs-frank-moss-was-a-symbol-of-nations-realignment.

2. Ralph Ellison, *Invisible Man* (New York: Random House, 1952).

3. Graham, *Presidents and the American Environment*, 279; Daynes and Sussman, *White House Politics and the Environment*, 174–75; Turner and Isenberg, *The Republican Reversal*, 60.

4. Edmund Morris, *Dutch: A Memoir of Ronald Reagan* (New York: Modern Library, 2000).

5. Graham, *Presidents and the American Environment*, 292–98; Turner and Isenberg, *The Republican Reversal*, 111–12.

6. Lindell L. Marsh, interview by Christopher E. Segal, November 30, 2018.

7. Joshua Chin, "Fighting to Save, and Popularize, San Bruno Mountain," *Bay Nature*, February 26, 2015, https://baynature.org/article/fighting-save-popularize-san-bruno-mountain.

8. Marsh, interview.

9. San Bruno Mountain Habitat Conservation Plan Steering Committee, *San Bruno Mountain Area Habitat Conservation Plan—Final* (November 1982), I-1.

10. Marsh, interview.

11. Endangered Species Act Amendments of 1982, H.R. Rep. No. 97–835, at 30–32 (1982) (Conf. Rep.); Lindell L. Marsh and Robert D. Thornton, "San Bruno Mountain Habitat Conservation Plan," in *Managing Land-Use Conflicts: Case Studies in Special Area Management*, ed. David J. Brower and Daniel S. Carol (Durham, NC: Duke University Press, 1987), 126–27.

12. Alan M. Gottlieb, ed., *The Wise Use Agenda: The Citizen's Policy Guide to Environmental Resource Issues* (Bellevue, WA: Free Enterprise Press, 1989); James Morton

Turner, "'The Specter of Environmentalism': Wilderness, Environmental Politics, and the Evolution of the New Right," *Journal of American History* 96, no. 1 (June 2009): 137–41.

13. Jaime Fuller, "The Long Fight between the Bundys and the Federal Government, from 1989 to Today," *Washington Post*, January 4, 2016, https://www.washingtonpost .com/news/the-fix/wp/2014/04/15/everything-you-need-to-know-about-the-long-fight -between-cliven-bundy-and-the-federal-government.

CHAPTER 4

1. Mark D. Ahner, *Can the United States Army Adjust to the Endangered Species Act of 1973?* (Carlisle Barracks, PA: U.S. Army War College, 1992), 15–18.

2. Ahner, *Can the United States Army Adjust to the Endangered Species Act of 1973?*, 1.

3. Human Protection Act of 1991, H.R. 3092, 102nd Cong. (1991); Balanced Economic and Environmental Priorities Act of 1991, H.R. 4058, 102nd Cong. (1991).

4. For example, the Endangered Species Act Procedural Reform Amendments of 1993, S. 1521, 103rd Cong. (1993).

5. Douglas P. Wheeler and Ryan M. Rowberry, "Habitat Conservation Plans and the Endangered Species Act," in *Endangered Species Act: Law, Policy, and Perspectives*, ed. Donald C. Baur and William Robert Irvin (Chicago: American Bar Association, 2010), 224. See also Bruce Babbitt, *Cities in the Wilderness: A New Vision of Land Use in America* (Washington, DC: Island Press, 2005), 62–63.

6. U. S. Fish and Wildlife Service, "West Fork Timber HCP (formerly Murray Pacific)," accessed April 23, 2021, https://ecos.fws.gov/ecp0/conservationPlan/plan?plan_id=116.

7. Brian Seasholes, *Fulfilling the Promise of the Endangered Species Act: The Case for an Endangered Species Reserve Program* (Reason Foundation, 2014), 42–44, https://reason .org/wp-content/uploads/files/endangered_species_act_reform.pdf.

8. Wheeler and Rowberry, "Habitat Conservation Plans and the Endangered Species Act," 224; Donald J. Barry, interview by Lowell E. Baier and Christopher E. Segal, January 25, 2019.

9. Seasholes, *Fulfilling the Promise of the Endangered Species Act*, 42–44.

10. Wheeler and Rowberry, "Habitat Conservation Plans and the Endangered Species Act," 224; Dianne K. Conway and Daniel S. Evans, "Salmon on the Brink: The Imperative of Integrating Environmental Standards and Review on an Ecosystem Scale," *Seattle University Law Review* 23 (2000): 991, https://digitalcommons.law.seattleu.edu/ cgi/viewcontent.cgi?article=1641&context=sulr; Barry, interview, January 25, 2019. These HCPs vary but can run for up to fifty to seventy-five years and provide regulatory certainty for not only landowners but also their successors-in-interest, as they can be filed as restrictive covenants or conservation easements that run with the land even when it is sold. U.S. Fish and Wildlife Service and National Marine Fisheries Service, *Habitat Conservation Planning and Incidental Take Permit Processing Handbook* (2016); Barry, interview, November 2, 2021.

11. U.S. Fish and Wildlife Service, "Environmental Conservation Online System: Conservation Plans by Type and Region," accessed August 7, 2023, https://ecos.fws .gov/ecp/report/conservation-plans-type-region; Wheeler and Rowberry, "Habitat Conservation Plans and the Endangered Species Act," 224–25; Michael J. Bean, *Landowner*

Assurances under the Endangered Species Act Working Paper (Madison, WI: Sand County Foundation/Environmental Policy Innovation Center, 2017), 9.

12. Barry, interview, January 25, 2019.

13. Barry, interview, January 25, 2019.

14. U.S. Fish and Wildlife Service, *Protecting America's Living Heritage: A Fair, Cooperative and Scientifically Sound Approach to Improving the Endangered Species Act* (March 6, 1995), http://www.fws.gov/policy/npi96_06.pdf.

15. The ten principles were as follows: (1) base ESA decisions on sound and objective science; (2) minimize social and economic impacts; (3) provide quick, responsive answers and certainty to landowners; (4) treat landowners fairly and with consideration; (5) create incentives for landowners to conserve species; (6) make effective use of limited public and private resources by focusing on groups of species dependent on the same habitat; (7) prevent species from becoming endangered or threatened; (8) promptly recover and delist threatened and endangered species; (9) promote efficiency and consistency; and (10) provide state, tribal, and local governments with opportunities to play a greater role in carrying out the ESA. U.S. Fish and Wildlife Service, *Protecting America's Living Heritage*, 3–4.

16. Barton H. Thompson Jr., "The Endangered Species Act: A Case Study in Takings & Incentives," *Stanford Law Review* 49, no. 2 (1997): 321; J. B. Ruhl, "While the Cat's Asleep: The Making of the 'New' ESA," *Natural Resources & Environment* 12, no. 3 (1998): 188; Donald J. Barry, interview by Christopher E. Segal and Lowell E. Baier, February 27, 2019.

17. Barry, interview, February 27, 2019.

18. U.S. Fish and Wildlife Service, *A Habitat Conservation Plan to Encourage the Voluntary Restoration and Enhancement of Habitat for the Red-Cockaded Woodpecker on Private and Certain Other Land in the Sandhills Region of North Carolina by Providing "Safe Harbor" to Participating Landowners* (1995), https://ecos.fws.gov/docs/plan_documents/tsha/tsha_451.pdf; David Drake and Edwin J. Jones, "Forest Management Decisions of North Carolina Landowners Relative to the Red-Cockaded Woodpecker," *Wildlife Society Bulletin* 30, no. 1 (2002): 121–30; Bruce Beard, interview by Christopher E. Segal, July 10, 2019.

19. Announcement of Final Safe Harbor Policy, 64 Fed. Reg. 32717 (June 17, 1999).

20. Drake and Jones, "Forest Management Decisions of North Carolina Landowners Relative to the Red-Cockaded Woodpecker"; Michael J. Bean, "Landowner Incentives and the Endangered Species Act," in Baur and Irvin, *Endangered Species Act*, 208–9.

21. Announcement of Final Safe Harbor Policy, 64 Fed. Reg. 32717, 32717.

22. U.S. Fish and Wildlife Service, "Environmental Conservation Online System"; Bean, *Endangered Species Act Safe Harbor Agreements*.

23. Bean, *Endangered Species Act Safe Harbor Agreements*, 4; U.S. Fish and Wildlife Service, "Environmental Conservation Online System: Red-Cockaded Woodpecker (*Picoides borealis*)," accessed October 3, 2018, https://ecos.fws.gov/ecp0/profile/speciesProfile?spcode=B04F.

24. Wendy Jackson, interview by Christopher E. Segal, January 28, 2019.

25. Announcement of Final Policy for Candidate Conservation Agreements with Assurances, 64 Fed. Reg. 32726 (June 17, 1999), 32735.

26. U.S. Fish and Wildlife Service, *Draft Candidate Conservation Agreements with Assurances Handbook* (June 2003), 14.

27. U.S. Fish and Wildlife Service, *Draft Candidate Conservation Agreements with Assurances Handbook*, 9–10.

28. U.S. Fish and Wildlife Service, "Environmental Conservation Online System."

29. Bean, *Landowner Assurances under the Endangered Species Act Working Paper*, 5. The 1999 Safe Harbor Policy recognized these as "a new category of 'enhancement of survival' permits." Announcement of Final Safe Harbor Policy, 64 Fed. Reg. 32717, 32722. Previously, enhancement of survival permits were issued for activities such as the removal of snail darters from the Little Tennessee River and release into the Hiwassee River back in the 1970s, for captive rearing of peregrine falcons and whooping cranes, for research on American alligators and gray wolves, and for zoo displays of exotic species, such as giant pandas, lemurs, and tigers. See, for example, Endangered Species Permit: Receipt of Addendum to Application, 41 Fed. Reg. 8515 (February 27, 1976); Endangered Species Permits: Official Action, 41 Fed. Reg. 9576 (March 5, 1976); and Endangered Species Permit: Notice of Receipt of Application, 39 Fed. Reg. 18483 (May 28, 1974).

30. George C. Edwards III, "Bill Clinton and His Crisis of Governance," *Presidential Studies Quarterly* 28, no. 4 (1998): 754–60.

31. Emergency Supplemental Appropriations and Rescissions for the Department of Defense to Preserve and Enhance Military Readiness Act of 1995, Pub. L. No. 104–6, 109 Stat. 73, 86 (1995); Benjamin Jesup, "Endless War or End This War? The History of Deadline Litigation under Section 4 of the Endangered Species Act and the Multidistrict Litigation Settlements," *Vermont Journal of Environmental Law* 14 (2013): 344–46.

32. Sandhya Somashekhar, "Gingrich Wild about Zoos," *New York Times*, December 9, 2011, https://www.washingtonpost.com/politics/2011/12/08/gIQAVb1yiO_story.html.

33. Michael J. Bean, "The Gingrich That Saved the ESA," *The Environmental Forum* 16, no. 1 (1999); Al Kamen, "Newt Gingrich's Animal Attraction Resurfaces," *Washington Post*, January 27, 2012, https://www.washingtonpost.com/blogs/in-the-loop/post/newt -gingrichs-animal-attraction-resurfaces/2012/01/26/gIQAofZhTQ_blog.html.

34. *Endangered Species Act: Washington, DC—Part III: Hearing on the Impact of the Endangered Species Act on the Nation before the Task Force on the Endangered Species Act of the House Committee on Resources*, 104th Cong. 20 (1995).

35. *Endangered Species Act: Washington, DC—Part III*, 20.

36. Kamen, "Newt Gingrich's Animal Attraction Resurfaces"; Bonnie B. Burgess, *Fate of the Wild: The Endangered Species Act and the Future of Biodiversity* (Athens: University of Georgia Press, 2001), 91–92.

37. Endangered Species Recovery Act of 1997, S. 1180, 105th Cong. (1997).

38. Brodie Farquhar, "Pombo Takes on the Endangered Species Act," *High Country News*, October 17, 2005, https://www.hcn.org/issues/308/15840/print_view.

39. Threatened and Endangered Species Recovery Act of 2005, H.R. 3824, 109th Cong. (2005).

40. *NWF's Kostyack and NCPPR's Ridenour Go Head to Head over House Species Legislation*, aired September 25, 2005, on E&ETV. *E&E News*, https://www.eenews.net/tv/videos/169/transcript.

41. "McNerney, Enviros Take Down Richard Pombo," *Capitol Weekly*, November 9, 2006, https://capitolweekly.net/mcnerney-enviros-take-down-richard-pombo; Betty Boxall, "Environmentalists Savor Pombo Defeat as Sign of Power," *Los Angeles Times*, November 9, 2006, https://www.latimes.com/archives/la-xpm-2006-nov-09-me -defenders9-story.html.

42. Rocky Barker, "Proposed Reform to Endangered Species Act Gets Cool Response," *Idaho Statesman*, August 17, 2008, http://bluefish.org/coolresp.htm.

43. Center for Biological Diversity, "Politics of Extinction, Attacks on the Endangered Species Act," accessed July 15, 2021, http://www.biologicaldiversity.org/compaigns/esa _attacks/table.html. A minimum of 600 total attacks were identified by using searches on Congress.gov to supplement the 2011–2020 data compiled by the Center for Biological Diversity.

CHAPTER 5

1. Nicholas Lemann, "No People Allowed," *The New Yorker*, November 22, 1999, https://www.newyorker.com/magazine/1999/11/22/no-people-allowed.

2. Lemann, "No People Allowed."

3. Arne Naess, "The Shallow and the Deep, Long-Range Ecology Movements: A Summary," *Inquiry* 16 (1973): 95–100.

4. Baier, *Inside the Equal Access to Justice Act*, 208–32.

5. National Animal Interest Alliance, "Quotes from the Leaders of the Animal Rights Movement," accessed August 8, 2010, http://www.naiaonline.org/body/article/archives/animalrightsquote.htm.

6. Tony Davis, "Firebrand Ways: A Visit with One of the Founders of the Center for Biological Diversity," *High Country News*, December 28, 2009, http://www.hcn.org/issues /41.22/firebrand-ways.

7. Annette McGivney, "Moses or Menace?" *Backpacker*, February 2003, 53; see also Edward Humes, *Eco Barons: The Dreamers, Schemers, and Millionaires Who Are Saving Our Planet* (New York: Ecco, 2009), 100; *ESA Decisions by Closed-Door Settlement: Short-Changing Science, Transparency, Private Property, and State & Local Economies; Legislative Hearing on H.R. 4315, H.R. 4316, H.R. 4317, and H.R. 4318 before the Committee on Natural Resources*, House of Representatives, 113th Cong. (2013), https://www .gpo.gov/fdsys/pkg/CHRG-113hhrg85958/pdf/CHRG-113hhrg85958.pdf (testimony of Kent Holsinger).

8. Jesup, "Endless War or End This War? The History of Deadline Litigation under Section 4 of the Endangered Species Act and the Multidistrict Litigation Settlements," 342–44.

9. In re: Endangered Species Act § 4 Deadline Litigation—MDL No. 2165, No. 1:2010-mc-00377-EGS (D.D.C. 2010).

10. *The Endangered Species Act: How Litigation Is Costing Jobs and Impeding True Recovery Efforts: Oversight Hearing before the House Committee on Natural Resources*, 112th Cong. 87 (2011) (written statement of Dan Ashe).

11. Revisions to the Regulations for Petitions, 81 Fed. Reg. 66462 (September 27, 2016).

12. Jacob Malcom, "Consequences of Resource Limitations on Endangered Species Act Implementation," in *Endangered Species Act: Law, Policy, and Perspectives*, ed. Donald C. Baur and Ya-Wei Li (Washington, DC: American Bar Association, 2021), 417–37.

13. U.S. Fish and Wildlife Service, "Environmental Conservation Online System: Listed Species Summary (Boxscore)," accessed December 13, 2023, https://ecos.fws.gov/ecp/report/boxscore; U.S. Fish and Wildlife Service, "Environmental Conservation Online System: Species Reports: Delisted Species," accessed December 13, 2023, https://ecos.fws.gov/ecp/report/species-delisted.

14. Daniel M. Ashe, interview by Lowell E. Baier and Christopher E. Segal, December 20, 2016.

15. U.S. Fish and Wildlife Service, "Environmental Conservation Online System: USFWS Threatened & Endangered Species Active Critical Habitat Report," accessed December 13, 2023, https://ecos.fws.gov/ecp/report/critical-habitat.

16. Michael Kiefer, "Owl See You in Court," *Phoenix New Times*, August 1, 1996; Proposed Determination of Critical Habitat for the Mexican Spotted Owl, 69 Fed. Reg. 63162 (December 7, 1994).

17. Houck, "The Endangered Species Act and Its Implementation by the U.S. Departments of Interior and Commerce," 277, 297.

18. Kiefer, "Owl See You in Court."

19. *Weyerhaeuser Co. v. U.S. Fish & Wildlife Service*, 139 S. Ct. 361 (2018).

20. Tate Watkins, "Endangered Frog's Survival Depends on Making Landowners Friends Not Foes," *Property and Environment Research Center*, September 27, 2018, https://www.perc.org/2018/09/27/endangered-frogs-survival-depends-on-making-landowners-friends-not-foes (quoting Edward Poitevent).

21. Threatened Species Status with Section 4(d) Rule for the Upper Coosa River Distinct Population Segment of Frecklebelly Madtom and Designation of Critical Habitat, 88 Fed. Reg. 13038 (March 2, 2023); U.S. Fish and Wildlife Service, *Recovery Outline for the Frecklebelly Madtom* (May 25, 2023), https://ecos.fws.gov/docs/recovery_plan/20230524_Frecklebelly%20madtom%20Recovery%20Outline.pdf.

22. Southeast Conservation Adaptation Strategy, "About SECAS," accessed February 23, 2021, https://secassoutheast.org/about.

23. Threatened Species Status with Section 4(d) Rule for the Upper Coosa River Distinct Population Segment of Frecklebelly Madtom and Designation of Critical Habitat, 85 Fed. Reg. 74050 (November 19, 2020), 74060.

24. House Committee on Merchant Marine and Fisheries, Endangered and Threatened Species Conservation Act of 1973, H.R. Rep. No. 93–412, at 4–5 (1973).

25. See, for example, Holly Doremus and Joel E. Pagel, "Why Listing May Be Forever: Perspectives on Delisting under the U.S. Endangered Species Act," *Conservation Biology* 15, no. 5 (October 2001): 1258–68; J. B. Ruhl, "The Battle over Endangered Species Act Methodology," *Environmental Law* 34, no. 2 (2004): 555–603.

26. Center for Biological Diversity, "Historic Accomplishment: Snail Darter Recovered," press release, August 31, 2021, https://biologicaldiversity.org/w/news/press-releases/historic-accomplishment-snail-darter-recovered-2021-08-31.

27. *Dialogue*, "Revisiting the Tennessee Story of the Snail Darter."

28. Revisions to the Regulations for Petitions, 81 Fed. Reg. 66462.

29. Stenographic Transcript of Hearings before the Committee on Merchant Marine and Fisheries, House of Representatives, 95th Congress, Markup Session on H.R. 13807 and H.R. 14104, September 19, 1978.

30. These nine species include certain populations of the American alligator, the Arctic peregrine falcon, the Aleutian Canada goose, the Tinian Monarch, the bald eagle, the Maguire daisy, the Oregon chub, the Okaloosa darter, and the snail darter.

31. Corbin Hiar, "Obama Admin Poised to Break Delisting Record," *Greenwire*, May 29, 2015, https://www.eenews.net/stories/1060019305; *Barriers to Endangered Species Act Delistings, Part II: Joint Hearing before the Subcommittee on the Interior and the Subcommittee on Health Care, Benefits and Administrative Rules of the House Committee on Oversight and Government Reform*, 114th Cong. 6 (2016) (statement of Dan Ashe).

32. Mary Ruckelshaus and Donna Darm, "Science and Implementation," in *The Endangered Species Act at Thirty: Conserving Biodiversity in Human-Dominated Landscapes*, vol. 2, ed. J. Michael Scott, Dale D. Goble, and Frank W. Davis (Washington, DC: Island Press, 2006), 122.

33. *Endangered Species Act Oversight: Hearings before the Subcommittee on Resource Protection of the Senate Committee on Environment and Public Works*, 37–40 (1977) (testimony of Frank Bond).

34. Ruckelshaus and Darm, "Science and Implementation," 122.

35. Dale D. Goble, "Recovery," in *Endangered Species Act: Law, Policy, and Perspectives*, ed. Donald C. Baur and William Robert Irvin (Chicago: American Bar Association, 2010), 87–90; J. Peyton Doub, *The Endangered Species Act: History, Implementation, Successes, and Controversies* (Boca Raton, FL: CRC Press, 2013), 122–23.

36. Jacob W. Malcom and Ya-Wei Li, "Missing, Delayed, and Old: The Status of ESA Recovery Plans," *Conservation Letters* 11, no. 6 (August 9, 2018), https://doi.org/10.1111/conl.12601.

37. U.S. Fish and Wildlife Service, "Environmental Conservation Online System: Listed Species Summary (Boxscore)." Note that unlike the Malcom and Li study, this is not limited to species listed at least 2.5 years ago.

38. Katherine Wright and Shawn Regan, "Missing the Mark: How the Endangered Species Act Falls Short of Its Own Recovery Goals," *Property and Environment Research Center*, July 26, 2023, https://www.perc.org/2023/07/26/missing-the-mark.

39. Final Rule to Remove the American Peregrine Falcon from the Federal List of Endangered and Threatened Wildlife, and to Remove the Similarity-of-Appearance Provision for Free-Flying Peregrines in the Conterminous United States, 64 Fed. Reg. 46542 (August 25, 1999), 46544–50.

40. Post-Delisting Monitoring Results for the American Peregrine Falcon (*Falco peregrinus anatum*), 2003, 71 Fed. Reg. 60563 (October 13, 2016).

41. Endangered Species Act Amendments of 1988, H.R. Rep. No. 100–928, at 21 (1988) (Conf. Rep.).

CHAPTER 6

1. National Park Service, *Yellowstone Wolf Project Biennial Report: 1995 and 1996*, 6, https://home.nps.gov/yell/learn/nature/upload/wolfrep95-96.pdf; Rachel Cohen, "25 Years Ago, Wolves Were Reintroduced to Idaho," *Boise State Public Radio*, January 14, 2020, https://www.boisestatepublicradio.org/news/2020-01-14/25-years-ago-wolves-were-reintroduced-to-idaho.

2. Cohen, "25 Years Ago, Wolves Were Reintroduced to Idaho"; Associated Press, "After Being Released in Idaho, Wolf Is Shot to Death on Ranch," *New York Times*, January 31, 1995.

3. National Park Service, "Yellowstone: Wolf Restoration," updated May 21, 2020, https://www.nps.gov/yell/learn/nature/wolf-restoration.htm.

4. U.S. Department of the Interior Office of Inspector General, "Report of Investigation: Julie MacDonald, Deputy Assistant Secretary, Fish, Wildlife and Parks," accessed September 22, 2023, https://grist.org/wp-content/uploads/2007/03/doi-ig-report_jm .pdf.

5. John M. Broder, "Ex-Interior Aide Is Sentenced in Lobbying Case," *New York Times*, June 27, 2007, https://www.nytimes.com/2007/06/27/washington/27griles.html.

6. Center for Biological Diversity, "Conservation Groups Challenge Kill-at-Will Policy for Wyoming Wolves," press release, September 10, 2012, https://www.biologicaldiversity .org/news/press_releases/2012/wolves-09-10-2012.html.

7. Wyo. H. J. No. 13, Resolution in Support of Central Park Wilderness (2012); Ben Yakas, "Wyoming Wants to Bring Wolves to Central Park," *Gothamist* (blog), *New York Public Radio*, February 12, 2012, https://gothamist.com/news/wyoming-wants-to-bring -wolves-to-central-park.

8. U.S. Fish and Wildlife Service, *Mexican Wolf Recovery Program: Progress Report #25* (2023), 16, https://www.fws.gov/sites/default/files/documents/Mexican-Wolf-2022 -Progress-Report-Final.pdf; Establishment of a Nonessential Experimental Population of the Mexican Gray Wolf in Arizona and New Mexico, 63 Fed. Reg. 1752 (January 12, 1998), 1768–70.

9. Lydia Warren, "Cages Built to Protect Kids from Wolves at New Mexico Bus Stops under Fire for 'Demonizing' the Endangered Animals," *The Daily Mail*, October 30, 2013, https://www.dailymail.co.uk/news/article-2479999/Cages-built-protect-kids -wolves-New-Mexico-bus-stops-demonizing-endangered-animals.html; International Wolf Center, "Are Wolves Dangerous to Humans?" accessed June 20, 2023, https://wolf .org/wolf-info/factsvsfiction/are-wolves-dangerous-to-humans.

10. Defenders of Wildlife, *Southwest Wolf Compensation Guidelines*, https://defenders .org/sites/default/files/publications/instructions_for_wolf_compensation_in_the_ southwest.pdf; Defenders of Wildlife, "Defenders Shifts Focus to Wolf Coexistence Partnerships," press release, August 20, 2010, https://defenders.org/newsroom/defenders -shifts-focus-wolf-coexistence-partnerships; Ralph Maughan, "Defenders Ends Wolf Depredation Payments," *The Wildlife News*, August 31, 2010.

11. U.S. Fish and Wildlife Service, "Wolf Livestock Loss Demonstration Project Grant Program," accessed June 20, 2023, https://www.fws.gov/service/wolf-livestock-loss -demonstration-project-grant-program.

12. Humane Society of the United States, *Government Data Confirm That Wolves Have a Negligible Effect on U.S. Cattle & Sheep Industries* (March 6, 2019), https://www .humanesociety.org/sites/default/files/docs/HSUS-Wolf-Livestock-6.Mar_.19Final.pdf; Living With Wolves, "Are Wolves Killing Lots of Cattle and Sheep?" accessed September 8, 2020, https://www.livingwithwolves.org/portfolio/are-wolves-killing-lots-of-cattle -and-sheep.

13. Living With Wolves, "Are Wolves Killing Lots of Cattle and Sheep?"

14. For example, John Williams, Douglas E. Johnson, Patrick E. Clark, Larry L. Larson, and Tyanne J. Roland, "Wolves—A Primer for Ranchers," EM 9142 (Oregon State University, March 2017), https://catalog.extension.oregonstate.edu/em9142/html, and Steve Stuebner, "Wolves Part 4: Unforeseen Impacts Caused by Wolves in Idaho," *Life on the Range*, accessed June 21, 2022, https://idrange.org/range-stories/north-central-idaho /unforeseen-impacts-caused-by-wolves-in-idaho.

15. Living With Wolves, "Are Wolves Killing Lots of Cattle and Sheep?"

16. Revision to the Regulations for the Nonessential Experimental Population of the Mexican Wolf; Final Rule, 80 Fed. Reg. 2512 (January 16, 2015); Lobos of the Southwest, "In the News: Revised Rules Let Mexican Wolves Roam Expanded Territory," accessed April 28, 2021, https://mexicanwolves.org/in-the-news-revised-rules-let -mexican-wolves-roam-expanded-territory.

17. *Center for Biological Diversity v. Jewell*, 2018 WL 1586651, 13 (D. Ariz. 2018).

18. Revision to the Nonessential Experimental Population of the Mexican Wolf, 86 Fed. Reg. 59953 (October 29, 2021).

19. Joe Duhownik, "Feds Defend Mexican Wolf Recovery Plan before Ninth Circuit Panel," *Courthouse News Service*, June 5, 2023, https://www.courthousenews.com/feds -defend-mexican-wolf-recovery-plan-before-ninth-circuit-panel.

20. For example, Christine Peterson, "25 Years after Returning to Yellowstone, Wolves Have Helped Stabilize the Ecosystem," *National Geographic*, July 10, 2020, https:// www.nationalgeographic.com/animals/article/yellowstone-wolves-reintroduction-helped -stabilize-ecosystem; Brodie Farquhar, "Wolf Reintroduction Changes Ecosystem in Yellowstone," *Outside*, June 30, 2021, https://www.yellowstonepark.com/things-to-do/ wildlife/wolf-reintroduction-changes-ecosystem; and Caeleigh MacNeil, "How Wolves Saved the Foxes, Mice and Rivers of Yellowstone National Park," *Earthjustice*, October 26, 2016, https://earthjustice.org/blog/2015-july/how-wolves-saved-the-foxes-mice-and -rivers-of-yellowstone-national-park.

21. For example, Defenders of Wildlife, "Gray Wolf," accessed September 8, 2020, https://defenders.org/wildlife/gray-wolf, and Center for Biological Diversity, "America's Gray Wolves: A Long Road to Recovery," accessed September 8, 2020, https://www .biologicaldiversity.org/campaigns/gray_wolves/index.html.

22. Douglas W. Smith, Rolf O. Peterson, Daniel R. MacNulty, and Michel Kohl, "The Big Scientific Debate: Trophic Cascades," *Yellowstone Science* 24, no. 1 (June 2016), https://www.nps.gov/articles/the-big-scientific-debate-trophic-cascades.htm. There is a persistent myth that wolves were reintroduced in order to control the elk population in anticipation of a trophic cascade, but this is incorrect. For example, BBC Future, "How

Reintroducing Wolves Helped Save a Famous Park," accessed September 8, 2020, https://www.bbc.com/future/article/20140128-how-wolves-saved-a-famous-park.

23. For example, Joel S. Brown, John W. Laundré, and Mahesh Gurung, "The Ecology of Fear: Optimal Foraging, Game Theory, and Trophic Interactions," *Journal of Mammalogy* 80, no. 2 (1999), https://academic.oup.com/jmammal/article/80/2/385/899792.

24. Emilene Ostlind and Dave Wade, "The Ecology of Fear: Elk Responses to Wolves in Yellowstone Are Not What We Thought," *Western Confluence*, January 6, 2014, https://westernconfluence.org/the-ecology-of-fear-elk-responses-to-wolves-in-yellowstone-are-not-what-we-thought; Michel T. Kohl, Daniel R. Stahler, Matthew C. Metz, James D. Forester, Matthew J. Kauffman, Nathan Varley, P. J. White, Douglas W. Smith, and Daniel R. MacNulty, "Diel Predator Activity Drives a Dynamic Landscape of Fear," *Ecological Monographs* 88, no. 4 (2018), https://esajournals.onlinelibrary.wiley.com/doi/full/10.1002/ecm.1313.

25. Wolf Conservation Center, "What Draws Tourists to Yellowstone? Wolves!" May 18, 2017, https://nywolf.org/2017/05/what-draws-tourists-to-yellowstone-wolves.

26. *Defenders of Wildlife v. Secretary, Department of the Interior*, 354 F.Supp.2d 1156 (D. Ore. 2005); *Defenders of Wildlife v. Hall*, 565 F.Supp.2d 1160 (D. Mont. 2008); *Defenders of Wildlife v. Salazar*, 729 F. Supp.2d 1207 (D. Mont. 2010).

27. Baier, *Inside the Equal Access to Justice Act*, 415, 457n15. The population's growth slowed but was not reversed.

28. For example, Keith Ridler, "Legislation Aims to Kill 90% of Wolves Roaming Idaho," *Associated Press*, April 20, 2021, https://apnews.com/article/wildlife-legislation-environment-wolves-bills-ccb1f49f74f0781ebbacd10c48f5abc1; Daniel M. Ashe, "Opinion: Cruelty toward Wolves Is Erasing Conservation Efforts. It's Time to Reinstate Their Protections.," *Washington Post*, August 3, 2021, https://www.washingtonpost.com/opinions/2021/08/03/wolves-idaho-montana-cruelty-conservation.

29. Center for Biological Diversity and the Humane Society of the United States, *Emergency Petition to Relist Gray Wolves (Canis lupus) in the Northern Rocky Mountains as an Endangered or Threatened "Distinct Population Segment" under the Endangered Species Act* (May 26, 2021), https://ecos.fws.gov/docs/tess/petition/992.pdf; Western Watersheds Project et al., *A Petition to List the Western North American Population of Gray Wolves (Canis lupus) as a Distinct Population Segment* (July 29, 2021), https://ecos.fws.gov/docs/tess/petition/3352.pdf.

30. 90-Day Finding for Two Petitions to List the Gray Wolf in the Western United States, 86 Fed. Reg. 51857 (September 17, 2021). The decision to initiate a status review was the only realistic option for the FWS. By planning to greatly reduce the number of wolves in their states, the two state legislatures were clearly violating the spirit—but not the letter—of the 2009 delisting rule that was in effect. Under the 1987 recovery plan, each state needed to maintain only 100 wolves, but prior to delisting, each had committed to maintaining 150, and populations were far higher. The new 2021 state laws called for reduction of their wolf populations, even all the way to the agreed-on limit of 150 in Montana, but still well above the original federal target of 100. So there is a strong states'-rights argument that the species should not be relisted. However, in both states, the directive to reduce wolf populations came from elected representatives, not

professional wildlife managers, which makes a mockery of the entire concept of professional, scientific management of wildlife resources by state governments. Furthermore, the 2009 delisting rule clearly stated that the FWS expected the states' approach to wolf management to continue and stated that a change in management would be a basis for a status review, which is what it has now initiated. So there is also a strong argument that the FWS could not turn a blind eye to the changes and do nothing. Moreover, there is a moral/philosophical question to consider regarding whether we should deliberately reduce the numbers of a population that took so much time, money, and labor to establish in the first place.

31. Removing the Gray Wolf (*Canis lupus*) from the List of Endangered and Threatened Wildlife, 85 Fed. Reg. 69778, 69790.

32. *Defenders of Wildlife v. Secretary, Department of the Interior, National Wildlife Federation v. Norton*, 386 F. Supp. 2d 553 (D. Vt. 2005); *Humane Society of the United States v. Kempthorne*, 579 F. Supp. 2d 7 (D.D.C. 2008); Reinstatement of Protections for the Gray Wolf in the Western Great Lakes in Compliance with Settlement Agreement and Court Order, 74 Fed. Reg. 47483 (September 16, 2009); *Humane Society of the United States v. Jewell*, 76 F. Supp. 3d 69 (D.D.C. 2014).

33. *Humane Society of the United States v. Zinke*, 865 F.3d 585 (D.C. Cir. 2017).

34. For example, *Humane Society of the United States v. Kempthorne*.

35. Removing the Gray Wolf (*Canis lupus*) from the List of Endangered and Threatened Wildlife, 85 Fed. Reg. 69778.

36. Michael Doyle, "Judge Restores Gray Wolf Protections," *E&E News PM*, February 10, 2022.

37. Associated Press, "Wisconsin Opens Early Wolf Hunt after Hunter Group Sued," February 22, 2021, https://www.mprnews.org/story/2021/02/22/wisconsin-opens-early -wolf-hunt-after-hunter-group-sued.

38. Christie Taylor, "Wisconsin Wolf Hunt Overshoots Quota, Worrying Conservationists," *Madison Public Radio*, March 5, 2021.

39. Neil Vigdor, "Wisconsin More Than Doubles Wolf-Hunting Quota, Angering Conservationists," *New York Times*, August 11, 2021, https://www.nytimes.com/2021 /08/11/us/wisconsin-wolves-hunting-kill.html; Jessie Opoien, "A Guide to the Legal Battles over Wisconsin's Wolf Hunt," *Capital Times*, September 4, 2021, https://madison .com/ct/news/local/govt-and-politics/election-matters/a-guide-to-the-legal-battles-over -wisconsin-s-wolf-hunt/article_d4b11d0b-d19e-5091-955f-c2a677fedc7c.html.

40. Danielle Kaeding, "Dane County Judge Temporarily Bars Wisconsin's Wolf Hunt, Orders DNR to Set Quota of Zero Wolves," *Wisconsin Public Radio*, October 22, 2021, https://www.wpr.org/dane-county-judge-temporarily-bars-wisconsins-wolf-hunt-orders -dnr-set-quota-zero-wolves.

41. For example, Cara Giaimo, "Yellowstone Owes Its Early Success to Public Bear Feeding," *Atlas Obscura*, August 31, 2016, https://www.atlasobscura.com/articles/ yellowstone-owes-its-early-success-to-public-bear-feeding.

42. National Park Service, "Yellowstone: Grizzly Bears & the Endangered Species Act," updated March 2, 2020, https://www.nps.gov/yell/learn/nature/bearesa.htm.

43. Scott Talbott, interview by Lowell E. Baier and Christopher E. Segal, July 9, 2017; Reuters, "Grizzlies in Yellowstone Area to Leave Endangered Species List," *Washington Post*, March 23, 2007.

44. Interagency Grizzly Bear Study Team, *Yellowstone Grizzly Bear Investigations 2022: Annual Report of the Interagency Grizzly Bear Study Team* (2023), 1, https://igbconline.org/wp-content/uploads/2023/09/Yellowstone-Grizzly-Bear-Investigations-2022-IGBST-Annual-Report.pdf.

45. Removing the Yellowstone Distinct Population Segment of Grizzly Bears from the Federal List of Endangered and Threatened Wildlife; 90-Day Finding on a Petition to List as Endangered the Yellowstone Distinct Population Segment of Grizzly Bears, 72 Fed. Reg. 14866 (March 29, 2007).

46. *Greater Yellowstone Coalition, Inc. v. Servheen*, 672 F. Supp. 2d 1105 (D. Mont. 2009).

47. *Greater Yellowstone Coalition, Inc. v. Servheen*, 665 F.3d 1015 (9th Cir. 2011). Although there were several other issues in the district court, the whitebark pine issue was the basis for the appellate decision.

48. Removing the Greater Yellowstone Ecosystem Population of Grizzly Bears from the Federal List of Endangered and Threatened Wildlife, 82 Fed. Reg. 30502 (June 30, 2017), 30536–40.

49. Removing the Greater Yellowstone Ecosystem Population of Grizzly Bears from the Federal List of Endangered and Threatened Wildlife, 82 Fed. Reg. 30502.

50. *Humane Society of the United States v. Zinke*.

51. Possible Effects of Court Decision on Grizzly Bear Recovery in the Conterminous United States, 82 Fed. Reg. 57698 (December 7, 2017); Review of 2017 Final Rule, Greater Yellowstone Ecosystem Grizzly Bears, 83 Fed. Reg. 18737 (April 30, 2018).

52. *Crow Indian Tribe v. United States*, 343 F. Supp. 3d 999 (D. Mont. 2018).

53. *Crow Indian Tribe v. United States*, 965 F.3d 662 (9th Cir. 2020).

54. Blackfoot Challenge, "History," accessed December 15, 2020, https://blackfootchallenge.org/history.

55. Seth M. Wilson, Elizabeth H. Bradley, and Gregory A. Neudecker, "Learning to Live with Wolves: Community-Based Conservation in the Blackfoot Valley of Montana," *Human–Wildlife Interactions* 11, no. 3 (Winter 2017): 247.

56. Wilson et al., "Learning to Live with Wolves," 247.

57. Jeff Welsch, "Keeping Montana . . . Montana," *Montana Quarterly*, Fall 2013, 11.

58. Wilson et al., "Learning to Live with Wolves," 247–48.

59. Wilson et al., "Learning to Live with Wolves," 248; Seth M. Wilson, Gregory A. Neudecker, and James J. Jonkel, "Human-Grizzly Bear Coexistence in the Blackfoot River Watershed, Montana: Getting Ahead of the Conflict Curve," in *Large Carnivore Conservation: Integrating Science and Policy in the North American West*, ed. Susan G. Clark and Murray B. Rutherford (Chicago: University of Chicago Press, 2014), 178–79, 183.

60. Wilson et al., "Learning to Live with Wolves," 248–50; Seth M. Wilson and Susan G. Clark, "Resolving Human-Grizzly Bear Conflict: An Integrated Approach in the Common Interest," in *Integrated Resource and Environmental Management: Concepts and Practice*, ed. D. Scott Slocombe and Kevin S. Hanna (Oxford: Oxford University Press,

2007), 155–56; Wilson et al., "Human-Grizzly Bear Coexistence in the Blackfoot River Watershed, Montana," 178–79, 184–90, 193–96. In addition, neighbors adopted bear-safe practices for disposal of household waste. Wilson et al., "Human-Grizzly Bear Coexistence in the Blackfoot River Watershed, Montana," 195.

61. Laura Lundquist, "NRCS Designs Grant for Blackfoot Challenge Grizzly Conflict Reduction," *Missoula Current*, November 10, 2021, https://www.kpax.com/news/western -montana-news/nrcs-designs-grant-for-blackfoot-challenge-grizzly-conflict-reduction.

62. Wilson et al., "Learning to Live with Wolves," 248–50; Wilson and Clark, "Resolving Human-Grizzly Bear Conflict," 155–56; Wilson et al., "Human-Grizzly Bear Coexistence in the Blackfoot River Watershed, Montana," 193–95; Rob Chaney, "Carcass Cleanup Tries to Stay Ahead of Grizzlies," *Missoulian*, April 29, 2019, https: //missoulian.com/news/local/carcass-cleanup-tries-to-stay-ahead-of-grizzlies/article _7b8cc949-8db0-5afa-98c0-f7fecb9dc32e.html.

63. Seth M. Wilson, *Summary of Grizzly and Wolf Conflicts for Blackfoot Watershed: Report to Montana Livestock Loss Board* (February 6, 2021).

64. Wilson et al., "Learning to Live with Wolves," 249; Wilson and Clark, "Resolving Human-Grizzly Bear Conflict," 157; Wilson, *Summary of Grizzly and Wolf Conflicts for Blackfoot Watershed*.

65. Wilson, *Summary of Grizzly and Wolf Conflicts for Blackfoot Watershed*.

66. Wilson and Clark, "Resolving Human-Grizzly Bear Conflict," 154–55.

67. Wilson et al., "Learning to Live with Wolves," 249; Wilson and Clark, "Resolving Human-Grizzly Bear Conflict: An Integrated Approach in the Common Interest," 154–57.

68. Wilson et al., "Learning to Live with Wolves," 253.

69. Wilson et al., "Learning to Live with Wolves," 250–53.

70. Wilson, *Summary of Grizzly and Wolf Conflicts for Blackfoot Watershed*.

71. Wilson et al., "Learning to Live with Wolves," 253–54; Wilson, *Summary of Grizzly and Wolf Conflicts for Blackfoot Watershed*.

72. Alaska Department of Fish and Game, "Wolf (*Canis lupus*) Species Profile," accessed November 3, 2021, https://www.adfg.alaska.gov/index.cfm?adfg=wolf.main; IUCN Red List, "Grey Wolf (*Canis lupus*)," accessed March 9, 2022, https://www .iucnredlist.org/species/pdf/133234888/attachment.

73. IUCN Red List, "Brown Bear (*Ursus arctos*)," June 21, 2023, https://www.iucnredlist .org/species/pdf/121229971/attachment.

74. Colorado Parks and Wildlife, "Gray and Mexican Wolves," June 21, 2023, https:// cpw.state.co.us/learn/Pages/SOC-Wolves.aspx.

75. Secretary of State, State of Colorado, *2020 Abstract of Votes Cast* (2020), 146, https://www .sos.state.co.us/pubs/elections/Results/Abstract/2020/2020BiennialAbstractBooklet.pdf.

76. For example, Jason Blevins, "Proposition 114 Explained: What's at Stake with the Effort to Reintroduce Gray Wolves in Colorado," *Colorado Sun*, September 24, 2020, https://coloradosun.com/2020/09/24/proposition-114-explained-wolf-reintroduction; "Proposition 114: The Reintroduction of Gray Wolves, Explained," *Colorado Public Radio*, October 20, 2020, https://www.cpr.org/2020/10/12/vg-2020-colorado-proposition-114 -reintroduction-of-gray-wolves-explained.

77. Colorado Parks and Wildlife, "Wolves in Colorado FAQ," accessed June 21, 2023, https://cpw.state.co.us/learn/Pages/Wolves-in-Colorado-FAQ.aspx.

CHAPTER 7

1. *Endangered Species: Hearings on H.R. 37, H.R. 470, H.R. 471, H.R. 1461, H.R. 1511, H.R. 2669, H.R. 2735, H.R. 3310, H.R. 3696, H.R. 3795, H.R. 4755, H.R. 2169, H.R. 4758, before the Subcommittee on Fisheries and Wildlife Conservation and the Environment of the House Committee on Merchant Marine and Fisheries,* 93rd Cong. 348 (1973) (hereinafter *House Subcommittee Hearings, 1973*).

2. Ya-Wei Li, "Recovery," in *Endangered Species Act: Law, Policy, and Perspectives,* ed. Donald C. Baur and Ya-Wei Li (Washington, DC: American Bar Association, 2021), 83.

3. Approximately every five years, the FWS publishes the National Survey of Fishing, Hunting, and Wildlife-Associated Recreation. The most recent survey, published in 2023, shows that U.S. residents over the age of sixteen took more than 1.7 billion trips in 2022 to participate in outdoor activities like fishing, hunting, wildlife watching, recreational boating, and target shooting. In doing so, outdoor enthusiasts spent an estimated $394 billion on equipment, travel, licenses, and fees last year alone. U.S. Fish and Wildlife Service, *2022 National Survey of Fishing, Hunting, and Wildlife-Associated Recreation,* September 2023, https://digitalmedia.fws.gov/digital/api/collection/document/id/2321/download. This is just one of many ways that sportsmen contribute to conservation. Another is through excise taxes collected by the federal government and then granted to state fish and wildlife agencies under the Pittman-Robertson and Dingell-Johnson Acts, which will be discussed in chapter 8. In 2022, these programs provided more than $1.6 billion in funding, which was provided to state agencies in 2023. To date, they have accounted for more than $42.8 billion (in inflation-adjusted 2022 dollars). See U.S. Fish and Wildlife Service, "Wildlife Restoration and Hunter Education Apportionments," *Wildlife and Sport Fish Restoration Program,* accessed March 7, 2023, https://tracs.fws.gov/wildlifeRestorationAndHunterEducationApportionments.html; U.S. Fish and Wildlife Service, "Sport Fish Restoration Apportionments," *Wildlife and Sport Fish Restoration Program,* accessed March 7, 2023, https://tracs.fws.gov/sportFishRestorationApportionments.html.

4. *House Subcommittee Hearings, 1973,* 241, 275–78, 280–81, 284–85, 297, 320, 335, 362, 378–79, 385.

5. *House Subcommittee Hearings, 1973,* 241.

6. Nathaniel P. Reed, interview by Lowell E. Baier and Christopher E. Segal, May 18, 2016; Dian Olson Belanger and Adrian Kinnane, *Managing American Wildlife: A History of the International Association of Fish and Wildlife Agencies* (Rockville, MD: Montrose Press, 2002), 49–53, 151–53, 171–76.

7. House Committee on Merchant Marine and Fisheries, Endangered and Threatened Species Conservation Act of 1973, H.R. Rep. No. 93–412 (1973), 6–7.

8. *Fish and Wildlife Miscellaneous, Part 1: Hearings on H.R. 2817, H.R. 4141, H.R. 4741, H.R. 3633, H.R. 8394, H.R. 7749, H.R. 4658, before the Subcommittee on Fisheries and Wildlife Conservation and the Environment of the House Committee on Merchant Marine and Fisheries,* 95th Cong. 94 (1977) (testimony of Lynn Greenwalt).

9. John D. Dingell Jr., interview by Lowell E. Baier, April, 29, 2016.

10. For example, Paul Phifer, presentation at the Endangered Species Act Luncheon Roundtable (Washington, DC), March 5, 2020.

11. U.S. Government Accountability Office, *Wildlife and Sport Fish Restoration: Competitive Grant Programs Managed Consistently with Relevant Regulations, but Monitoring Could Be Improved*, GAO-18–303 (February 2018), https://www.gao.gov/assets/gao-18 -303.pdf. As evidenced by the title of this report, the prevailing pressure is for even *more*, not *less*, federal oversight.

12. "Journals of the Lewis & Clark Expedition, March 2, 1806," accessed August 11, 2017, https://lewisandclarkjournals.unl.edu/item/lc.jrn.1806-03-02#n28030201.

13. Indicator species are also sometimes referred to as "umbrella" species. For an in-depth discussion of many other sagebrush-associated species, see generally T. E. Remington, P. A. Deibert, S. E. Hanser, D. M. Davis, L. A. Robb, and J. L. Welty, *Sagebrush Conservation Strategy—Challenges to Sagebrush Conservation*, U.S. Geological Survey Open-File Report 2020–1125 (2021), https://doi.org/10.3133/ofr20201125.

14. National Wildlife Federation, "Saving the Greater Sage-Grouse," accessed May 7, 2018, https://www.nwf.org/Our-Work/Wildlife-Conservation/Greater-Sage-Grouse.

15. William Temple Hornaday, *Save the Sage Grouse from Extinction: A Demand from Civilization to the Western States*, Permanent Wild Life Protection Fund Bulletin 5 (New York: New York Zoological Park, 1916); Michael R. Conover and Anthony J. Roberts, "Declining Populations of Greater Sage-Grouse: Where and Why," *Human–Wildlife Interactions* 10, no. 2 (2016): 217.

16. U.S. Fish and Wildlife Service, "Who's on the Lek: A Guide to Players," accessed August 11, 2017, https://www.fws.gov/mountain-prairie/species/birds/sagegrouse/ Primer2SGWhoOnTheLek.pdf.

17. Today, there is limited hunting of greater sage-grouse in those areas where the population is large enough to support it.

18. John W. Connelly, Steven T. Knick, Michael A. Schroeder, and San J. Stiver, *Conservation Assessment of Greater Sage-Grouse and Sagebrush Habitats* (Boise, ID: Western Association of Fish and Wildlife Agencies, 2004).

19. "A Timeline of Sage Grouse Conservation," *Western Confluence*, 2013, http://www .sagegrouseinitiative.com/wp-content/uploads/2014/03/Western-%0AConfluence-»-A -timeline-of-sage-grouse-conservation.pdf; U.S. Fish and Wildlife Service, "Who's on the Lek."

20. S. J. Stiver, A. D. Apa, J. R. Bohne, S. D. Bunnell, P. A. Deibert, S. C. Gardner, M. A. Hilliard, C. W. McCarthy, and M. A. Schroeder, *Greater Sage-Grouse Comprehensive Conservation Strategy* (Cheyenne, WY: Western Association of Fish and Wildlife Agencies, 2006), https://wdfw.wa.gov/sites/default/files/publications/01317/wdfw01317.pdf; San Stiver, interview by Christopher E. Segal, March 25, 2014. The conservation strategy contemplated funding sage-grouse conservation through the proposed "North American Sagebrush Ecosystem Conservation Act" (NASECA), modeled on the North American Waterfowl Management Plan and its funding mechanism, the North American Wetlands Conservation Act. NASECA never became law, but sage-grouse conservation proceeded with other funding sources.

21. Memorandum of Understanding among Western Association of Fish and Wildlife Agencies and U.S. Department of Agriculture, Forest Service and U.S. Department of the Interior, Bureau of Land Management and U.S. Department of the Interior, Fish and Wildlife Service (2008), https://www2.usgs.gov/mou/docs/mou_sage_grouse.pdf. This agreement established two committees to help facilitate the implementation of the 2006 document: the technical "Range-Wide Interagency Sage-Grouse Conservation Team" and the advisory "Executive Oversight Committee." The oversight committee was charged with gathering information on sage-grouse programs, providing those programs with a range-wide perspective, and helping implement conservation strategies. The conservation team was directed to provide technical support and did so by publishing the "Near-Term Greater Sage-Grouse Conservation Action Plan" in 2012. San Stiver, interview by Christopher E. Segal, March 25, 2014; Range-Wide Interagency Sage-Grouse Conservation Team, *Near-Term Greater Sage-Grouse Conservation Action Plan* (September 11, 2012), http://sagemap.wr.usgs.gov/docs/rs/NTSGConservation%20Action%20Plan .pdf.

22. "A Timeline of Sage Grouse Conservation"; *Defining Species Conservation Success: Tribal, State and Local Stewardship vs. Federal Courtroom Battles and Sue-and-Settle Practices: Oversight Hearing before the House Committee on Natural Resources*, 113th Cong, 33–37 (2013) (testimony of Steve Ferrell); Kenneth E. Mayer, interview by Christopher E. Segal, March 24, 2014.

23. San Stiver, "Background on the Executive Oversight Committee," presentation to the Sagebrush Executive Oversight Committee, Association of Fish and Wildlife Agencies Annual Meeting (St. Paul, MN), September 24, 2019. Funding came from the FWS (including the Partners for Fish and Wildlife program), the BLM, the U.S. Forest Service, the federal Farm Bill (through the NRCS's Working Lands for Wildlife program), all of the states in the species' range, private landowners and local groups, and others.

24. Harney County Sage Grouse CCAA Steering Committee, "2011 Year in Review," *Harney County Sage Grouse CCAA Steering Committee Quarterly Update*, January 2012, 3, https://www.co.harney.or.us/PDF_Files/Sage%20Grouse/Harney%20County%20Sage -grouse%20newsletter%20volume%201%20issue%201%20January%202012.pdf.

25. For example, Harney County Sage Grouse CCAA Steering Committee, "2011 Year in Review." In addition, the committee worked to communicate with local landowners and inform them of other developments in sage-grouse conservation as the BLM and the Forest Service worked on their land management plans and the FWS moved toward the 2015 listing deadline.

26. *Greater Sage-Grouse Programmatic Candidate Conservation Agreement with Assurances for Private Rangelands in Harney County, Oregon between the Harney Soil and Water Conservation District and the United States Fish and Wildlife Service* (April 25, 2014), https: //www.fws.gov/oregonfwo/toolsforlandowners/habitatconservationplans/Documents/5 .13.14%20Harney%20SWCD%20CCAA-Final.pdf.

27. *Greater Sage-Grouse Programmatic Candidate Conservation Agreement with Assurances for Private Rangelands in Harney County, Oregon between the Harney Soil and Water Conservation District and the United States Fish and Wildlife Service*; Oregon Department of Fish and Wildlife, *Greater Sage-Grouse Conservation Assessment and Strategy for*

Oregon: A Plan to Maintain and Enhance Populations and Habitat (April 22, 2011), https://www.dfw.state.or.us/wildlife/sagegrouse/docs/20110422_GRSG_April_Final%2052511.pdf.

28. Public Lands Foundation, "Landscape Stewardship Awards—2014," September 9, 2014, https://publicland.org/awards/harney-county-soil-and-water-conservation-district; Daniel M. Ashe, "Keynote Address," presentation at the 81st North American Wildlife and Natural Resources Conference (Pittsburgh, PA), March 15, 2016.

29. 12-Month Finding on a Petition to List Greater Sage-Grouse (*Centrocercus urophasianus*) as an Endangered or Threatened Species, 80 Fed. Reg. 59858 (October 2, 2015).

30. Eric Mortenson, "States Adopt Plans as Sage Grouse Listing Decision Approaches," *Capital Press*, April 6, 2015, updated December 13, 2018, https://www.capitalpress.com/nation_world/nation/states-adopt-plans-as-sage-grouse-listing-decision-approaches/article_ba299df4-38e5-5cd9-811c-60359aa7947a.html.

31. Removing the Kirtland's Warbler from the Federal List of Endangered and Threatened Wildlife, 84 Fed. Reg. 54436 (October 9, 2019), 54444.

32. Removing the Kirtland's Warbler from the Federal List of Endangered and Threatened Wildlife, 84 Fed. Reg. 54436.

33. See generally Lowell E. Baier, *Saving Species on Private Lands: Unlocking Incentives to Conserve Wildlife and Their Habitats* (Lanham, MD: Rowman & Littlefield, 2020), 73–126.

34. Michael Martinez, "Working Lands for Wildlife: Targeted Landscape-Scale Wildlife Habitat Conservation," *Natural Resources & Environment* 29, no. 3 (2015): 37.

35. Natural Resources Conservation Service, *Working Lands for Wildlife Obligations, All Years All Species* (2019); Natural Resources Conservation Servce, "Working Lands for Wildlife," accessed June 5, 2023, https://www.wlfw.org.

36. U.S. Department of the Interior, *Budget Justifications and Performance Information Fiscal Year 2021: Fish and Wildlife Service* (2021), HC-1, https://www.fws.gov/budget/2021/FY2021-FWS-Budget-Justification.pdf.

37. U.S. Environmental Protection Agency, "Wetlands Compensatory Mitigation," accessed October 4, 2019, https://www.epa.gov/sites/production/files/2015-08/documents/compensatory_mitigation_factsheet.pdf.

38. Compensatory Mitigation for Losses of Aquatic Resources, 73 Fed. Reg. 19594 (April 10, 2008).

39. Wayne Walker, interview by Christopher E. Segal, October 5, 2018.

40. Endangered Species Act Compensatory Mitigation Policy, 81 Fed. Reg. 95316 (December 27, 2016), 95316.

41. U.S. Fish and Wildlife Service Mitigation Policy, 83 Fed. Reg. 36472 (July 30, 2018); Endangered Species Act Compensatory Mitigation Policy, 83 Fed. Reg. 36469 (July 30, 2018).

42. U.S. Department of the Interior, Secretarial Order No. 3349: American Energy Independence (March 29, 2017); U.S. Department of the Interior, Secretarial Order No. 3360: Rescinding Authorities Inconsistent with Secretary's Order 3349, "American Energy Independence" (December 22, 2017); Executive Order No. 13,783: Promoting Energy Independence and Economic Growth, 82 Fed. Reg. 16093 (March 28, 2017);

U.S. Fish and Wildlife Service Mitigation Policy, 83 Fed. Reg. 36472; Endangered Species Act Compensatory Mitigation Policy, 83 Fed. Reg. 36469. In addition, under the Trump administration, the BLM concluded that it could not require compensatory mitigation at all, and the Forest Service concluded that while it could require compensatory mitigation, it could not require net conservation benefits and could require only no net loss. Bureau of Land Management, Instruction Memorandum 2018–093: Compensatory Mitigation (July 24, 2018). See also U.S. Forest Service, *Greater Sage-Grouse Proposed Land Management Plan Amendments (LMPA) and Draft Environmental Impact Statement (DEIS) for the Intermountain and Rocky Mountain Regions* (October 2018): ES-3, https://www.fs.usda.gov/Internet/FSE_DOCUMENTS/fseprd598215.pdf ("Net conservation gain changed to no net loss of habitat . . ."). The conflicts between the Obama and Trump approaches to compensatory mitigation were significant because the changes made by the Obama administration were based on the best contemporary science. By adopting a net conservation benefit standard, the administration worked to refine compensatory mitigation so that it could be used to effectively conserve entire landscapes and ecosystems. James R. Lyons, interview by Lowell E. Baier and Christopher E. Segal, October 9, 2018.

43. Michael J. Bean, *Habitat Exchanges: A New Tool to Engage Landowners in Conservation Working Paper* (Madison, WI: Sand County Foundation/Environmental Policy Innovation Center, 2017), 67.

44. U.S. Fish and Wildlife Service Mitigation Policy and Endangered Species Act Compensatory Mitigation Policy, 88 Fed. Reg. 31000 (May 15, 2023).

45. William M. (Mac) Thornberry National Defense Authorization Act for Fiscal Year 2021, § 329, Pub. L. No. 116–283, 134 Stat. 3388, 3527 (2021); Compensatory Mitigation Mechanisms, 87 Fed. Reg. 45076 (July 27, 2022).

46. U.S. Fish and Wildlife Service, *Guidance for the Establishment, Use, and Operation of Conservation Banks* (May 2, 2003).

47. Aldo Leopold, "Conservation Economics," in *The River of the Mother of God and Other Essays by Aldo Leopold*, ed. Susan L. Flader and J. Baird Callicott (Madison: University of Wisconsin Press, 1991), 200.

CHAPTER 8

1. Association of Fish and Wildlife Agencies, *The State Conservation Machine* (2017), 3, http://www.fishwildlife.org/files/The_State_Conservation_Machine-FINAL.pdf.

2. U.S. Fish and Wildlife Service, "Wildlife Restoration and Hunter Education Apportionments"; U.S. Fish and Wildlife Service, "Sport Fish Restoration Apportionments."

3. Department of the Interior and Related Agencies Appropriations Act, 2002, Pub. L. No. 107–63, 115 Stat. 414, 422 (2001).

4. U.S. Fish and Wildlife Service, "State Wildlife Grant Program Apportionment History," 2019, https://wsfrprograms.fws.gov/Subpages/GrantPrograms/SWG/SWGAppt2002-2019.pdf; U.S. Fish and Wildlife Service, "State Wildlife Grant Program—2020 Apportionment," https://www.fws.gov/wsfrprograms/Subpages/GrantPrograms/SWG/SWG2020Apportionment.pdf; U.S. Fish and Wildlife Service, "State Wildlife Grant Program—2021 Apportionment," https://www.fws.gov/wsfrprograms/Subpages/GrantPrograms/SWG/SWG2021Apportionment.pdf. SWG

funding requires states to develop State Wildlife Action Plans that identify species of greatest conservation need and other wildlife conservation priorities. Best practices for these plans were developed by the Association of Fish and Wildlife Agencies and have been regularly updated. Association of Fish and Wildlife Agencies, *Best Practices for State Wildlife Action Plans: Voluntary Guidance to States for Revision and Implementation* (November 2012), https://www.fishwildlife.org/application/files/3215/1856/0300/SWAP_Best_Practices_Report_Nov_2012.pdf.

5. U.S. Fish and Wildlife Service, "Endangered Species Act: Grants: Overview," accessed November 17, 2021, https://www.fws.gov/endangered/grants.

6. U.S. Department of the Interior, *Budget Justifications and Performance Information Fiscal Year 2021: Fish and Wildlife Service*, BG-2.

7. U.S. Department of the Interior, *Budget Justifications and Performance Information Fiscal Year 2021: Fish and Wildlife Service*, BG-2–BG-3.

8. U.S. Department of the Interior, *Budget Justifications and Performance Information Fiscal Year 2021: Fish and Wildlife Service*, ES-12, MB-3.

9. Partners for Fish and Wildlife Act, Pub. L. No. 109–294, 120 Stat. 1351 (2006).

10. U.S. Department of the Interior, *Budget Justifications and Performance Information Fiscal Year 2021: Fish and Wildlife Service*, HC-1.

11. Gary Frazer, e-mail to Lowell E. Baier, June 23, 2021.

12. For example, Paul Phifer, presentation at the Endangered Species Act Luncheon Roundtable.

13. Baier, *Saving Species on Private Lands*, 131–38.

14. For more information on the National Fish and Wildlife Foundation, see Baier, *Saving Species on Private Lands*, 154–56.

15. National Fish and Wildlife Foundation, *2022 Conservation Investments*, 2, https://www.nfwf.org/sites/default/files/2023-03/nfwf-2022-conservation-investments-online.pdf.

16. Baier, *Saving Species on Private Lands*, 139–42.

17. Baier, *Saving Species on Private Lands*, 131.

18. Baier, *Saving Species on Private Lands*, 143–44.

19. Baier, *Saving Species on Private Lands*, 144–45.

20. National Audubon Society, "12 Ways the Inflation Reduction Act Will Benefit Birds and People," August 17, 2022, https://www.audubon.org/news/12-ways-inflation-reduction-act-will-benefit-birds-and-people.

21. U.S. Fish and Wildlife Service, "Bipartisan Infrastructure Law Funds Proven Projects for Wildlife," accessed July 11, 2023, https://www.fws.gov/initiative/directors-priorities/bipartisan-infrastructure-law-funds-proven-projects-wildlife.

22. Timothy Male, "Making an Asset of Endangered Species Recovery," in *The Codex of the Endangered Species Act, Volume II: The Next Fifty Years*, ed. Lowell E. Baier, John F. Organ, and Christopher E. Segal (Lanham, MD: Rowman & Littlefield, 2023), 14–29.

23. Douglas P. Wheeler and Dale Ratliff, "The Future of Section 10—Multiple Species Habitat Conservation Plans?" in Baier et al., *The Codex of the Endangered Species Act, Volume II*, 187–211.

24. James L. Cummins, "Reauthorizing a New Farm Bill," *Fair Chase*, Summer 2023, 8.

25. Cummins, "Reauthorizing a New Farm Bill," 8.

26. Cummins, "Reauthorizing a New Farm Bill," 8.

27. Cummins, "Reauthorizing a New Farm Bill," 8–9.

28. James L. Cummins, interview by Christopher E. Segal, March 6, 2019.

29. More comprehensive information can be found in Baier, *Saving Species on Private Lands*.

30. Cummins, "Reauthorizing a New Farm Bill," 9–10.

31. Cummins, "Reauthorizing a New Farm Bill," 10.

CHAPTER 9

1. Wayne Walker, interview by Christopher E. Segal, October 5, 2018.

2. Walker, interview; Donald "Pete" Gober, interview by Lowell E. Baier and Christopher E. Segal, May 25, 2018.

3. Environmental Policy Innovation Center, "A Guide to the Revised Endangered Species Regulations," accessed July 6, 2021, http://policyinnovation.org/esaregs19.

4. Revision of Regulations for Interagency Cooperation, 88 Fed. Reg. 40753 (June 22, 2023).

5. Enhancement of Survival and Incidental Take Permits, 88 Fed. Reg. 8380 (February 9, 2023).

6. Wheeler and Ratliff, "The Future of Section 10—Multiple Species Habitat Conservation Plans?"

7. This approach was recently proposed by three independent studies: Temple Stoellinger, Michael Brennan, Sara Brodnax, Ya-Wei Li, Murray Feldman, and Bob Budd, "Improving Cooperative State and Federal Species Conservation Efforts," *Wyoming Law Review* 20, no. 1 (2020): Article 3; Fischman, Meretsky, and Castelli, "Collaborative Governance under the Endangered Species Act"; and David Willms, "Unlocking the Full Power of Section 4(d) to Facilitate Collaboration and Greater Species Recovery," in Baier et al., *The Codex of the Endangered Species Act, Volume II*, 30–50.

8. Jonathan Wood and Tate Watkins, "The Future of Habitat: Lessons from the Dusky Gopher Frog Conflict," in Baier et al., *The Codex of the Endangered Species Act, Volume II*, 103–24.

9. This discussion is based on Male, "Making an Asset of Endangered Species Recovery." Male also notes that as a fail-safe, the Services should be able to suspend section 7(a)(1) plans that are not being implemented and initiate section 7(a)(2) consultation for ongoing actions previously covered by any such plan. This way, the proposed section 7(a)(1) approach would not imperil any species.

10. 12-Month Finding on a Petition to List the New England Cottontail as an Endangered or Threatened Species, 80 Fed. Reg. 55286 (September 15, 2015). In this case, habitat loss was due to commercial, industrial, and residential development but also reforestation over the past 100 years. Moreover, the New England cottontail had actually been a candidate for listing between 1982 and 1996 as a Category 2 species, meaning there was a lack of data on its biological vulnerability. In 1996, the FWS stopped including Category 2 species on the candidate list.

11. 12-Month Finding on a Petition to List the New England Cottontail as an Endangered or Threatened Species, 80 Fed. Reg. 55286.

12. Anthony Tur, e-mail to Christopher E. Segal, February 17, 2021.

13. Wendi Weber, interview by Christopher E. Segal and Lowell E. Baier, January 21, 2021.

14. Weber, interview.

15. Weber, interview.

16. Weber, interview.

17. Steven Fuller, Anthony Tur, and New England Cottontail Technical Committee, *Conservation Strategy for the New England Cottontail (*Sylvilagus transitionalis*)* (November 20, 2012), https://newenglandcottontail.org/sites/default/files/conservation_strategy_final_12-3-12.pdf.

18. Fuller et al., *Conservation Strategy for the New England Cottontail (Sylvilagus transitionalis)*, 1.

19. Weber, interview.

20. Weber, interview.

21. Roger Williams Park Zoo, "New England Cottontail," accessed January 28, 2021, https://www.rwpzoo.org/conservation-local/new-england-cottontail.

22. Louis Perrotti, interview by Christopher E. Segal, January 29, 2021.

23. Weber, interview.

24. Weber, interview.

25. Weber, interview; Working Together for the New England Cottontail, "Partners," accessed January 28, 2021, https://newenglandcottontail.org/partners.

26. Wendi Weber, "ESA Helps Write New Chapter for Peter Cottontail," *Huffington Post*, September 22, 2015, https://newenglandcottontail.org/news/esa-helps-write-new-chapter-peter-cottontail.

CHAPTER 10

1. Center for Biological Diversity, *Petition to List 404 Aquatic, Riparian and Wetland Species from the Southeastern United States as Threatened or Endangered under the Endangered Species Act* (April 20, 2010), https://www.fws.gov/southeast/pdf/petition/404-aquatic.pdf; John Platt, "Petition Filed to Protect 404 Southeastern U.S. Species," *Scientific American*, April 22, 2010, https://blogs.scientificamerican.com/extinction-countdown/petition-filed-to-protect-404-southeastern-u-s-species.

2. Southeast Conservation Adaptation Strategy, "About SECAS," accessed February 23, 2021, https://secassoutheast.org/about.

3. Jimmy Bullock, interview by Christopher E. Segal, January 26, 2021.

4. Removal of the Louisiana Black Bear from the Federal List of Endangered and Threatened Wildlife and Removal of Similarity-of-Appearance Protections for the American Black Bear, 81 Fed. Reg. 13124 (March 11, 2016).

5. Memorandum of Understanding among U.S. Fish and Wildlife Service, National Alliance of Forest Owners, and National Council for Air and Stream Improvement, Inc. (2023), https://nafoalliance.org/wp-content/uploads/2023/03/Wildlife-Conservation-Initiative-MOU-USFWS-NAFO-NCASI.pdf.

6. Dohner, personal communication, as described in James F. Bullock Jr. and Cynthia K. Dohner, "Conservation Without Conflict: A Collaborative Approach to the Endangered Species Act," in Baier et al., *The Codex of the Endangered Species Act, Volume II*, 250–67, 260. The roots, development, and future potential of Conservation Without Conflict are recounted in detail by Bullock and Dohner, two of the leaders and participants in that history. The discussion here is based largely on their work.

7. Bullock and Dohner, "Conservation Without Conflict," 255.

8. Cindy K. Dohner and Wendi Weber, Conservation Without Conflict, 2017, as described in Bullock and Dohner, "Conservation Without Conflict," 260.

9. Cindy K. Dohner, Wendi Weber, James Cummins, and Jimmy F. Bullock Jr., Conservation Without Conflict purpose statement, as quoted in Bullock and Dohner, "Conservation Without Conflict," 261.

10. John F. Organ, "Conservation Without Conflict: Saving Species and Livelihoods," *Fair Chase*, Summer 2020, https://www.boone-crockett.org/conservation-without-conflict-saving-species-and-livelihoods.

11. U.S. Department of the Interior, U.S. Department of Agriculture, U.S. Department of Commerce, and Council on Environmental Quality, *Conserving and Restoring America the Beautiful*, 2021, https://www.doi.gov/sites/doi.gov/files/report-conserving-and-restoring-america-the-beautiful-2021.pdf.

12. The Nature Conservancy, "Water Funds Toolbox," accessed February 26, 2021, https://waterfundstoolbox.org.

13. Conor P. McGowan, Nathan Allan, and David R. Smith, "The Species Status Assessment: A Framework for Assessing Species Status and Risk to Support Endangered Species Management Decisions," in Baier et al., *The Codex of the Endangered Species Act, Volume II*, 87–102; Temple Stoellinger, Michael Brennan, Sara Brodnax, Ya-Wei Li, Murray Feldman, and Bob Budd, "Improving Cooperative State and Federal Species Conservation Efforts," in Baier et al., *The Codex of the Endangered Species Act, Volume II*, 212–37.

14. Ken Elowe, "The ESA and Landscape Conservation: A Vision for the Future," in Baier et al., *The Codex of the Endangered Species Act, Volume II*, 125–37.

15. "Resource Management Service, LLC," accessed November 19, 2021, https://resourcemgt.com.

16. See, for example, Future Learn, "What is Corporate Sustainability and Why Is It Important?" September 10, 2021, https://www.futurelearn.com/info/blog/what-is-corporate-sustainability; Tima Bansal and Devika Agarwal, "Corporate Sustainability—Meaning, Examples and Importance," *Network for Business Sustainability*, March 10, 2021, https://www.nbs.net/articles/corporate-sustainability-meaning-examples-and-importance; Yale University Library, "Sustainability: Corporate Sustainability," accessed November 19, 2021, https://guides.library.yale.edu/c.php?g=296179&p=2582471.

17. Patagonia, "Environmental Activism," accessed November 19, 2021, https://www.patagonia.com/activism.

18. Ben & Jerry's, "How Ben & Jerry's Is Fighting Global Warming," February 24, 2015, https://www.benjerry.com/values/issues-we-care-about/climate-justice/fighting-global-warming.

19. For example, The Nature Conservancy, "Companies Investing in Nature," accessed November 19, 2021, https://www.nature.org/en-us/about-us/who-we-are/how-we-work/working-with-companies/companies-investing-in-nature1; Ecosystem Marketplace, "Private Investment in Conservation Reaches $8.2 Billion," press release, January 11, 2017, https://www.ecosystemmarketplace.com/articles/private-investment-in-conservation-reaches-8-2-billion; Ben Schiller, "Investors Are Pouring Money into Conservation Efforts (It's Actually a Great Investment)," *Fast Company*, February 2, 2017, https://www.fastcompany.com/3067575/investors-are-pouring-money-into-conservation-efforts-its-actually-a-; World Economic Forum, "3 Reasons Companies are Investing in Forest Conservation and Restoration, and How They Do It," accessed November 19, 2021, https://www.weforum.org/agenda/2021/06/3-reasons-companies-are-investing-in-forest-conservation-and-restoration-and-how-they-do-it.

20. For example, Tim Quinson and Mathieu Benhamou, "Banks Always Backed Fossil Fuel over Green Projects—Until This Year," *Bloomberg Green*, May 19, 2021, https://www.bloomberg.com/graphics/2021-wall-street-banks-ranked-green-projects-fossil-fuels.

CHAPTER 11

1. Aldo Leopold, "The Ecological Conscience," in *The River of the Mother of God and Other Essays by Aldo Leopold*, ed. Susan L. Flader and J. Baird Callicott, 338–46 (Madison: University of Wisconsin Press, 1991), 346.

2. Greenwald et al., "Extinction and the U.S. Endangered Species Act."

3. International Union for Conservation of Nature, "IUCN Red List of Threatened Species," accessed December 7, 2023, https://www.iucnredlist.org/search.

4. For example, Simon Worrall, "Saving Half the Planet for Nature Isn't as Crazy as It Seems," *National Geographic*, March 27, 2016, https://www.nationalgeographic.com/adventure/article/160327-wilson-half-planet-conservation-climate-change-extinction-ngbooktalk (interview with Edward O. Wilson). Most of these species would be insects and other small, geographically isolated species unknown to science. For example, studies of arthropods in South American rainforests typically turn up thousands of individual species per acre and tens or hundreds of thousands of individuals, including many new species. See, for example, World Wildlife Fund, "How Many Species Are We Losing?" accessed October 27, 2021, https://wwf.panda.org/discover/our_focus/biodiversity/biodiversity, and Jeremy Hance, "The Great Insect Dying: The Tropics in Trouble and Some Hope," *Mongabay*, June 10, 2019, https://news.mongabay.com/2019/06/the-great-insect-dying-the-tropics-in-trouble-and-some-hope. One 2022 study estimated that there may be as many as 9,000 species of trees that we've never identified. Marc Heller, "Thousands of Tree Species Go Undiscovered, Scientists Say," *Greenwire*, February 3, 2022.

5. Edward O. Wilson, "The Current State of Biological Diversity," in *Biodiversity*, ed. E. O. Wilson (Washington, DC: National Academies Press, 1988), 13.

6. For example, Katy Daigle and Julia Janicki, "Extinction Crisis Puts 1 Million Species on the Brink," *Reuters*, December 23, 2022, https://www.reuters.com/lifestyle/science/extinction-crisis-puts-1-million-species-brink-2022-12-23.

7. World Wildlife Fund, *Living Planet Report 2020: Bending the Curve of Biodiversity Loss* (2020), 6, https://f.hubspotusercontent20.net/hubfs/4783129/LPR/PDFs/ENGLISH-FULL.pdf (emphasis added). Losses were a staggering 94 percent in tropical subregions in the Americas and 84 percent for global freshwater habitats. World Wildlife Fund, *Living Planet Report 2020*, 6, 24.

8. For example, Sophie Lewis, "Animal Populations Worldwide Have Declined Nearly 70% in Just 50 Years, New Report Says," *CBS News*, September 10, 2020, https://www.cbsnews.com/news/endangered-species-animal-population-decline-world-wildlife-fund-new-report, and Nathan Rott, "The World Lost Two-Thirds of Its Wildlife in 50 Years. We Are to Blame," *NPR*, September 10, 2020, https://www.npr.org/2020/09/10/911500907/the-world-lost-two-thirds-of-its-wildlife-in-50-years-we-are-to-blame. Some media coverage of the WWF's Living Planet Reports has correctly cautioned that the top-line summary of a "70 percent loss" of species' populations (or 60 percent in 2018) may be an overstatement, as we must ask 70 percent of *what* species? For example, small percentage changes in the population of very numerous species can make the situation seem worse than it is. Moreover, the WWF reports are limited to data on a relatively small number of species of mammals, birds, reptiles, amphibians, and fish that make up the WWF's Living Planet Index. Specifically, the index tracks 16,700 populations of 4,000 species. So it is an extrapolation to say that we have lost 70 percent of *all* mammals, birds, reptiles, amphibians, and fish. And it is a leap to imply that we have lost so many of the invertebrate species, which vastly outnumber vertebrates and are entirely omitted from the WWF's statistics. Nonetheless, even critical media outlets emphasize that the situation is clearly very dire. See, for example, Ed Yong, "Wait, Have We Really Wiped Out 60 Percent of Animals?" *The Atlantic*, October 31, 2018, https://www.theatlantic.com/science/archive/2018/10/have-we-really-killed-60-percent-animals-1970/574549, and Elizabeth Anne Brown, "Widely Misinterpreted Report Still Shows Catastrophic Animal Decline," *National Geographic*, November 1, 2018, https://www.nationalgeographic.com/animals/article/animal-decline-living-planet-report-conservation-news.

9. World Wildlife Fund, "Living Planet Report 2020," accessed October 28, 2021, https://livingplanet.panda.org/en-us.

10. Lesli Allison, interview by Christopher E. Segal, February 17, 2017.

11. "Dramatic Decline in Western Butterfly Populations Linked to Fall Warming," *ScienceDaily*, March 4, 2021, https://www.sciencedaily.com/releases/2021/03/210304145405.htm; M. L. Forister, C. A. Halsch, C. C. Nice, J. A. Fordyce, T. E. Dilts, J. C. Oliver, K. L. Prudic, et al. "Fewer Butterflies Seen by Community Scientists across the Warming and Drying Landscapes of the American West." *Science* 371, no. 6533 (March 5, 2021): 1042–45, https://doi.org/10.1126/science.abe5585; Xerces Society, "Monarchs in Decline," accessed September 18, 2023, https://xerces.org/monarchs/conservation-efforts; Liz Kimbrough, "Western Monarch Populations Reach Highest Number in Decades," *Mongabay*, January 31, 2023, https://news.mongabay.com/2023/01/monarch-populations-rebound-but-its-still-a-long-journey-to-recovery.

12. 12-Month Finding for the Monarch Butterfly, 85 Fed. Reg. 81813 (December 17, 2020), https://www.govinfo.gov/content/pkg/FR-2020-12-17/pdf/2020-27523.pdf.

13. Aaron Sidder, "New Map Highlights Bee Population Declines across the U.S.," *Smithsonian*, February 23, 2017, https://www.smithsonianmag.com/smart-news/new-map-highlights-bee-population-declines-across-us-180962268; Siobhán Dunphy, "Declining Wild Bee Populations Threaten Key Food Crops," *European Scientist*, March 8, 2020, https://www.europeanscientist.com/en/agriculture/declining-wild-bee-populations-threaten-key-food-crops; University of New Hampshire, "More Than Dozen Wild Bee Species Declining in Northeastern U.S.," *ScienceDaily*, April 9, 2019, https://www.sciencedaily.com/releases/2019/04/190409093801.htm; Minna E. Mathiasson and Sandra M. Rehan, "Status Changes in the Wild Bees of North-Eastern North America over 125 Years Revealed through Museum Specimens," *Insect Conservation and Diversity* 12, no. 4 (2019), https://doi.org/10.1111/icad.12347; Eduardo E. Zattara, "Historic Data Hint at a Worldwide Decline of Wild Bee Biodiversity," *Indiana University Bloomington*, January 22, 2021, https://biology.indiana.edu/news-events/news/2021/wild-bee-decline.html; Eduardo E. Zattara and Marcelo A. Aizen, "Worldwide Occurrence Records Suggest a Global Decline in Bee Species Richness," *One Earth* 4, no. 1 (2021), https://doi.org/10.1016/j.oneear.2020.12.005; Catrin Einhorn and Nadja Popovich, "This Map Shows Where Biodiversity Is Most at Risk in America," *New York Times*, March 3, 2022, https://www.nytimes.com/interactive/2022/03/03/climate/biodiversity-map.html; Kelsey Kopec and Lori Ann Burd, *Pollinators in Peril, Center for Biological Diversity*, February 2017, https://www.biologicaldiversity.org/campaigns/native_pollinators/pdfs/Pollinators_in_Peril.pdf, 1; Romina Rader, Ignasi Bartomeus, Lucas A. Garibaldi, Michael P. D. Garratt, Brad G. Howlett, Rachael Winfree, Saul A. Cunningham, et al., "Non-Bee Insects Are Important Contributors to Global Crop Pollination," *PNAS* 113, no. 1 (2016), https://doi.org/10.1073/pnas.1517092112; Xerces Society, "Who Are the Pollinators?" accessed November 11, 2021, https://www.xerces.org/pollinator-conservation/about-pollinators; U.S. Forest Service, "Bird Pollination," accessed November 11, 2021, https://www.fs.fed.us/wildflowers/pollinators/animals/birds.shtml; U.S. Forest Service, "Bat Pollination," accessed November 11, 2021, https://www.fs.fed.us/wildflowers/pollinators/animals/bats.shtml.

14. Nathalie Steinhauer, "United States Honey Bee Colony Losses 2022–23: Preliminary Results from the Bee Informed Partnership," *Bee Informed*, June 22, 2023, https://beeinformed.org/2023/06/22/united-states-honey-bee-colony-losses-2022-23-preliminary-results-from-the-bee-informed-partnership. These commercial honeybees are bred and then moved around the country to pollinate crops such as almonds, apples, blueberries, cherries, pumpkins, and many others, including fruits, nuts, vegetables, legumes, oilseeds, and forage crops.

15. Josh Woods, "US Beekeepers Continue to Report High Colony Loss Rates, No Clear Progression toward Improvement," Auburn University, updated June 25, 2021, June 24, 2021, https://ocm.auburn.edu/newsroom/news_articles/2021/06/241121-honey-bee-annual-loss-survey-results.php; Robert M. Nowierski, "Pollinators at a Crossroads," *U.S. Department of Agriculture*, July 29, 2021, https://www.usda.gov/media/blog/2020/06/24/pollinators-crossroads.

16. See Hamilton, "The Role of Biodiversity in a Resilient America."

17. Aaron Sidder, "New Map Highlights Bee Population Declines across the U.S."

18. For example, *Alabama–Tombigbee River Coalition v. Kempthorne*, 477 F.3d at 1273–74.

19. Make Way for Monarchs, "Michelle Obama Creates First-Ever 'Pollinator Garden' at the White House," April 30, 2014, https://makewayformonarchs.org/2014/04/michelle-obama-creates-first-ever-pollinator-garden-at-the-white-house; Juliet Eilperin, "Michelle Obama Makes the Pitch for Pollinators," *Washington Post*, June 3, 2015, https://www.washingtonpost.com/news/post-politics/wp/2015/06/03/michelle-obama-makes-the-pitch-for-pollinators.

20. Environmental Protection Agency, "Federal Pollinator Health Task Force: EPA's Role," updated January 19, 2017, https://19january2017snapshot.epa.gov/pollinator-protection/federal-pollinator-health-task-force-epas-role_.html. Although the goals of the strategy were not met, it helped to establish pollinator conservation as a major issue. These goals were to restore honeybee colony health to sustainable levels by 2025, increase eastern monarch butterfly populations to 225 million butterflies by 2020, and restore or enhance 7 million acres of land for pollinators over five years. Another important event for pollinator conservation was the 2017 listing of the rusty patched bumblebee as an endangered species, the first species of bumblebee to ever be listed under the ESA. Endangered Species Status for Rusty Patched Bumble Bee, 82 Fed. Reg. 3186 (January 11, 2017).

21. Natural Resources Conservation Servce, "Monarch Butterflies," accessed July 19, 2023, https://www.nrcs.usda.gov/programs-initiatives/monarch-butterflies.

22. National Wildlife Federation, "Mayors' Monarch Pledge: Program Overview," accessed July 18, 2023, https://www.nwf.org/MayorsMonarchPledge/About/Overview; Patrick Fitzgerald, *Monarch Conservation in America's Cities* (National Wildlife Federation, 2015), http://www.nwf.org/~/media/PDFs/Garden-for-Wildlife/Monarch-Conservation-in-Americas-Cities_Guide-121715.pdf.

23. U.S. Fish and Wildlife Service, *Nationwide Candidate Conservation Agreement for Monarch Butterfly on Energy and Transportation Lands: An integrated Candidate Conservation Agreement with Assurances (CCAA) and Candidate Conservation Agreement (CCA)* (March 2020), https://www.fws.gov/savethemonarch/pdfs/Final_CCAA_040720_Fully%20Executed.pdf; Tim Male, interview by Lowell E. Baier and Christopher E. Segal, March 5, 2020.

24. "UIC Leads Largest Nationwide Effort to Protect the Monarch Butterfly," *UIC Today*, April 8, 2020, https://today.uic.edu/uic-leads-largest-nationwide-effort-to-protect-the-monarch-butterfly; Timothy Male, "Saving the Monarch Butterfly," *Sand County Foundation*, April 9, 2020, https://sandcountyfoundation.org/news/2020/saving-the-monarch-butterfly; Male, interview.

25. Commission for Environmental Cooperation, *North American Monarch Conservation Plan* (2008), https://www.fs.fed.us/wildflowers/pollinators/Monarch_Butterfly/news/documents/Monarch-Monarca-Monarque.pdf.

26. Midwest Association of Fish and Wildlife Agencies, *Mid-America Monarch Conservation Strategy* (2018), http://www.mafwa.org/wp-content/uploads/2018/05/MidAmericaMonarchStrategyDraft_May11_2018.pdf.

27. Midwest Association of Fish and Wildlife Agencies, *Mid-America Monarch Conservation Strategy 2018–2038*, http://www.mafwa.org/wp-content/uploads/2019/02 /MAMCS_Summary_Handout_web.pdf; Midwest Association of Fish and Wildlife Agencies, "Q&A Messages for Mid-America Monarch Conservation Strategy," accessed July 19, 2023, http://www.mafwa.org/?page_id=2414.

28. Western Association of Fish and Widlife Agencies, "Western Monarch Butterfly Conservation Plan, 2019–2069," accessed July 19, 2023, https://wafwa.org/wpdm -package/western-monarch-butterfly-conservation-plan-2019-2069; Xerces Society, "Western Monarch Conservation," accessed July 19, 2023, https://xerces.org/monarchs /western-monarch-conservation.

29. Monarch Joint Venture, *2021 Monarch Conservation Implementation Plan* (2021), https://monarchjointventure.org/images/uploads/documents/2021_Monarch _Conservation_Implementation_Plan.pdf.

30. U.S. Fish and Wildlife Service, "Monarch Conservation Database," accessed November 12, 2021, https://storymaps.arcgis.com/stories/8169fcb99632492cb738716 54b9310bb; Habitat Conservation Assistance Network, "Monarch Butterfly," accessed March 8, 2022, https://www.habitatcan.org/Monarch-Butterfly.

31. For example, John W. Reid and Thomas E. Lovejoy, *Ever Green: Saving Big Forests to Save the Planet* (New York: W. W. Norton & Company, 2022); Lee Hannah and Thomas E. Lovejoy, eds., *Biodiversity and Climate Change: Transforming the Biosphere* (New Haven, CT: Yale University Press, 2019); Thomas E. Lovejoy and Lee Hannah, eds., *Climate Change and Biodiversity* (New Haven, CT: Yale University Press, 2005).

32. For example, Andrew Revkin, "E.O. Wilson at 90: The Conservation Legend Shares Dreams for the Future," *National Geographic*, June 10, 2019, https://www .nationalgeographic.com/environment/article/eo-wilson-conservation-legend-90-save -space-for-nature-save-planet; Michael Becker, "MSU Presents Presidential Medal to Famed Scientist Edward O. Wilson," *MSU News Service*, April 9, 2009, https://www .montana.edu/news/7071/msu-presents-presidential-medal-to-famed-scientist-edward -o-wilson. The concept of biodiversity is relatively modern, with its popularization the result of work done in the early 1980s by scientists such as Lovejoy, Wilson, Bruce A. Wilcox, and W. G. Rosen. Wilson, together with Robert MacArthur, developed the theory of island biogeography in the 1960s. The theory states that larger habitats support larger and more diverse populations and species, which in turn are less prone to extinction. Mathematical equations define the application of the theory to areas of various sizes. Applying these to the entire Earth leads to the conclusion that half of the planet should be set aside as intact natural areas. See, for example, Half-Earth Project, "Discover Half-Earth," accessed October 28, 2021, https://www.half-earthproject.org/discover-half -earth, and Revkin, "E.O. Wilson at 90." This became a major cause for Dr. Wilson, and the basis for his 2016 book *Half-Earth*. Edward O. Wilson, *Half-Earth* (New York: Liveright, 2016). His conclusion is that if we can protect half of Earth, we can preserve 85 to 90 percent of species. Revkin, "E.O. Wilson at 90." The same math formed the basis for "Nature Needs Half," a campaign launched by the WILD Foundation in 2009, as well as numerous other studies and papers. See, for example, Nature Needs Half, "History," accessed October 28, 2021, https://natureneedshalf.org/who-we-are/history.

33. Edward O. Wilson, "Introduction," in Tony Hiss, *Rescuing the Planet: Protecting Half the Land to Heal the Earth* (New York: Alfred A. Knopf, 2021). The more recent "Thirty by Thirty" initiative is another response to these crises. See, for example, "Saving Nature Will Take Bold Action: 'Thirty by Thirty,'" *Natural Resources Defense Council*, February 7, 2020, https://www.nrdc.org/experts/alison-chase/saving-nature-will-take-bold -action-thirty-thirty.

AFTERWORD

1. Carson, *Silent Spring*, 277.

2. Fischman, Meretsky, and Castelli, "Collaborative Governance under the Endangered Species Act," 978.

Bibliography

Books

Ahner, Mark D. *Can the United States Army Adjust to the Endangered Species Act of 1973?* Carlisle Barracks, PA: U.S. Army War College, 1992.

Babbitt, Bruce. *Cities in the Wilderness: A New Vision of Land Use in America.* Washington, DC: Island Press, 2005.

Baier, Lowell E. *The Codex of the Endangered Species Act, Volume I: The First Fifty Years.* With Christopher E. Segal and a foreword by Douglas Brinkley. Lanham, MD: Rowman & Littlefield, 2023.

———. *Inside the Equal Access to Justice Act: Environmental Litigation and the Crippling Battle over America's Lands, Endangered Species, and Critical Habitats.* Foreword by John D. Dingell Jr. Lanham, MD: Rowman & Littlefield, 2016.

———. *Saving Species on Private Lands: Unlocking Incentives to Conserve Wildlife and Their Habitats.* With Christopher E. Segal and a foreword by Dirk Kempthorne. Lanham, MD: Rowman & Littlefield, 2020.

Belanger, Dian Olson, and Adrian Kinnane. *Managing American Wildlife: A History of the International Association of Fish and Wildlife Agencies.* Rockville, MD: Montrose Press, 2002.

Brinkley, Douglas. *Silent Spring Revolution: John F. Kennedy, Rachel Carson, Lyndon Johnson, Richard Nixon, and the Great Environmental Awakening.* New York: Harper, 2022.

Burgess, Bonnie B. *Fate of the Wild: The Endangered Species Act and the Future of Biodiversity.* Athens: University of Georgia Press, 2001.

Carson, Rachel. *Silent Spring.* Boston: Houghton Mifflin, 1962.

Chase, Alston. *In a Dark Wood: The Fight over Forests and the Rising Tyranny of Ecology.* New York: Houghton Mifflin, 1995.

Commoner, Barry. *The Closing Circle: Nature, Man, and Technology.* New York: Alfred A. Knopf, 1971.

Daynes, Byron W., and Glen Sussman. *White House Politics and the Environment: Franklin D. Roosevelt to George W. Bush.* College Station: Texas A&M University Press, 2010.

Doub, J. Peyton. *The Endangered Species Act: History, Implementation, Successes, and Controversies.* Boca Raton, FL: CRC Press, 2013.

Drew, Elizabeth. *Richard M. Nixon.* New York: Time Books/Henry Holt, 2007.

Ehrlichman, John. *Witness to Power: The Nixon Years.* New York: Simon and Schuster, 1982.

Ellison, Ralph. *Invisible Man*. New York: Random House, 1952.

Flippen, J. Brooks. *Nixon and the Environment*. Albuquerque: University of New Mexico Press, 2000.

Gottlieb, Alan M., ed. *The Wise Use Agenda: The Citizen's Policy Guide to Environmental Resource Issues*. Bellevue, WA: Free Enterprise Press, 1989.

Graham, Mary. *The Morning after Earth Day*. Washington, DC: Brookings Institution, 1999.

Graham, Otis L., Jr. *Presidents and the American Environment*. Lawrence: University Press of Kansas, 2015.

Hall, H. Dale. *Compelled: From the Yazoo Pumps to Polar Bears and Back: The Evolution of Natural Resource Conservation and Law*. Memphis, TN: Ducks Unlimited, Inc., 2022.

Hannah, Lee, and Thomas E. Lovejoy, eds. *Biodiversity and Climate Change: Transforming the Biosphere*. New Haven, CT: Yale University Press, 2019.

Hays, Samuel P. *A History of Environmental Politics since 1945*. Pittsburgh, PA: University of Pittsburgh Press, 2000.

Hornaday, William Temple. *Save the Sage Grouse from Extinction: A Demand from Civilization to the Western States*. Permanent Wild Life Protection Fund Bulletin 5. New York: New York Zoological Park, 1916.

Humes, Edward. *Eco Barons: The Dreamers, Schemers, and Millionaires Who Are Saving Our Planet*. New York: Ecco, 2009.

Layzer, Judith A. *Open for Business: Conservatives' Opposition to Environmental Regulation*. Cambridge, MA: MIT Press, 2012.

Lovejoy, Thomas E., and Lee Hannah, eds. *Climate Change and Biodiversity*. New Haven, CT: Yale University Press, 2005.

Mann, Charles C., and Mark L. Plummer. *Noah's Choice: The Future of Endangered Species*. New York: Alfred A. Knopf, 1995.

Marcot, Bruce G., and Jack Ward Thomas. *Of Spotted Owls, Old Growth, and New Policies: A History since the Interagency Scientific Committee Report*. General Technical Report PNW-GTR-408. Portland, OR: U.S. Department of Agriculture, Forest Service, Pacific Northwest Research Station, 1997.

Montrie, Chad. *The Myth of Silent Spring: Rethinking the Origins of American Environmentalism*. Berkeley: University of California Press, 2018.

———. *A People's History of Environmentalism in the United States*. New York: Continuum Press, 2011.

Morris, Edmund. *Dutch: A Memoir of Ronald Reagan*. New York: Modern Library, 2000.

Murchison, Kenneth M. *The Snail Darter Case: TVA versus the Endangered Species Act*. Lawrence: University Press of Kansas, 2007.

Patterson, James T. *Restless Giant: The United States from Watergate to Bush v. Gore*. Oxford: Oxford University Press, 2005.

Pearlstein, Rick. *Nixonland*. New York: Scribner, 2008.

Rathlesberger, James, ed. *Nixon and the Environment: The Politics of Devastation*. New York: Village Voice, 1972.

Reed, Nathaniel Pryor. *Travels on the Green Highway: An Environmentalist's Journey.* Hobe Sound: Reed Publishing Company, 2016.

Reid, John W., and Thomas E. Lovejoy. *Ever Green: Saving Big Forests to Save the Planet.* New York: W. W. Norton, 2022.

Tober, James A. *Wildlife and the Public Interest: Nonprofit Organizations and Federal Wildife Policy.* New York: Praeger, 1989.

Train, Russell E. *Politics, Pollution and Pandas.* Washington, DC: Island Press, 2003.

Turner, James Morton, and Andrew C. Isenberg. *The Republican Reversal: Conservatives and the Environment from Nixon to Trump.* Cambridge, MA: Harvard University Press, 2018.

Whitaker, John C. *Striking a Balance: Environment and Natural Resources Policy in the Nixon-Ford Years.* Washington, DC: AEI-Hoover Policy Studies, 1976.

Wilson, Edward O. *The Diversity of Life.* Cambridge, MA: Harvard University Press, 1992.

———. *Half-Earth.* New York: Liveright, 2016.

Yaffee, Steven Lewis. *Prohibitive Policy: Implementing the Federal Endangered Species Act.* Cambridge, MA: MIT Press, 1982.

JOURNAL ARTICLES AND BOOK CHAPTERS

"Address by Lee M. Talbot, Senior Scientist, Council on Environmental Quality, upon Receipt of the Schweitzer Centenary Medal." *Animal Welfare Institute Information Report* 24, no. 1 (January–March 1975).

Bean, Michael J. "The Gingrich That Saved the ESA." *The Environmental Forum* 16, no. 1 (1999).

———. "Landowner Incentives and the Endangered Species Act." In *Endangered Species Act: Law, Policy, and Perspectives*, edited by Donald C. Baur and William Robert Irvin, 206–18. Chicago: American Bar Association, 2010.

Brown, Joel S., John W. Laundré, and Mahesh Gurung. "The Ecology of Fear: Optimal Foraging, Game Theory, and Trophic Interactions." *Journal of Mammalogy* 80, no. 2 (1999): 385–99. https://academic.oup.com/jmammal/article/80/2/385/899792.

Bullock, James F., Jr., and Cynthia K. Dohner. "Conservation Without Conflict: A Collaborative Approach to the Endangered Species Act." In *The Codex of the Endangered Species Act, Volume II: The Next Fifty Years*, edited by Lowell E. Baier, John F. Organ, and Christopher E. Segal, 250–67. Lanham, MD: Rowman & Littlefield, 2023.

Conover, Michael R., and Anthony J. Roberts. "Declining Populations of Greater Sage-Grouse: Where and Why." *Human–Wildlife Interactions* 10, no. 2 (2016): 217–29.

Conway, Dianne K., and Daniel S. Evans. "Salmon on the Brink: The Imperative of Integrating Environmental Standards and Review on an Ecosystem Scale." *Seattle University Law Review* 23 (2000): 977–1018. https://digitalcommons.law.seattleu.edu/cgi/viewcontent.cgi?article=1641&context=sulr.

Cummins, James L. "Reauthorizing a New Farm Bill." *Fair Chase*, Summer 2023, 8–10.

Doremus, Holly, and Joel E. Pagel. "Why Listing May Be Forever: Perspectives on Delisting under the U.S. Endangered Species Act." *Conservation Biology* 15, no. 5 (October 2001): 1258–68.

Dorn, Trilby C. E. "Logging without Laws: The 1995 Salvage Logging Rider Radically Changes Policy and the Rule of Law in the Forests." *Tulane Environmental Law Journal* 9, no. 2 (1996): 447–82.

Drake, David, and Edwin J. Jones. "Forest Management Decisions of North Carolina Landowners Relative to the Red-Cockaded Woodpecker." *Wildlife Society Bulletin* 30, no. 1 (2002): 121–30.

Edwards, George C., III. "Bill Clinton and His Crisis of Governance." *Presidential Studies Quarterly* 28, no. 4 (1998): 754–60.

Elowe, Ken. "The ESA and Landscape Conservation: A Vision for the Future." In *The Codex of the Endangered Species Act, Volume II: The Next Fifty Years*, edited by Lowell E. Baier, John F. Organ, and Christopher E. Segal, 125–37. Lanham, MD: Rowman & Littlefield, 2023.

Fischman, Robert L., Vicky J. Meretsky, and Matthew P. Castelli. "Collaborative Governance under the Endangered Species Act: An Empirical Analysis of Protective Regulations." *Yale Journal on Regulation* 38 (2021): 976–1058.

Flournoy, Alyson C. "Beyond the Spotted Owl Problem: Learning from the Old-Growth Controversy." *Harvard Environmental Law Review* 17 (1993): 261–332.

Forister, M. L., C. A. Halsch, C. C. Nice, J. A. Fordyce, T. E. Dilts, J. C. Oliver, K. L. Prudic, et al. "Fewer Butterflies Seen by Community Scientists across the Warming and Drying Landscapes of the American West." *Science* 371, no. 6533 (March 5, 2021): 1042–45. https://doi.org/10.1126/science.abe5585.

Goble, Dale D. "Recovery." In *Endangered Species Act: Law, Policy, and Perspectives*, edited by Donald C. Baur and William Robert Irvin, 70–103. Chicago: American Bar Association, 2010.

Greenwald, Noah, Kieran F. Suckling, Brett Hartl, and Loyal A. Mehrhoff. "Extinction and the U.S. Endangered Species Act." *PeerJ* 7 (2019): e6803.https://pubmed.ncbi.nlm.nih.gov/31065461.

Hebblewhite, Mark, Clifford A. White, Clifford G. Nietvelt, John A. McKenzie, Tomas E. Hurd, John M. Fryxell, Suzanne E. Bayley, and Paul C. Paquet. "Human Activity Mediates a Trophic Cascade Caused by Wolves." *Ecology* 86, no. 8 (August 2005): 2135–44.

Houck, Oliver A. "The Endangered Species Act and Its Implementation by the U.S. Departments of Interior and Commerce." *University of Colorado Law Review* 64 (1993): 277–370.

Jesup, Benjamin. "Endless War or End This War? The History of Deadline Litigation under Section 4 of the Endangered Species Act and the Multidistrict Litigation Settlements." *Vermont Journal of Environmental Law* 14 (2013): 327–87.

Kohl, Michel T., Daniel R. Stahler, Matthew C. Metz, James D. Forester, Matthew J. Kauffman, Nathan Varley, P. J. White, Douglas W. Smith, and Daniel R. MacNulty. "Diel Predator Activity Drives a Dynamic Landscape of Fear."

Ecological Monographs 88, no. 4 (2018): 638–52. https://esajournals.onlinelibrary .wiley.com/doi/full/10.1002/ecm.1313.

Leopold, Aldo. "Conservation Economics." In *The River of the Mother of God and Other Essays by Aldo Leopold*, edited by Susan L. Flader and J. Baird Callicott, 193–202. Madison: University of Wisconsin Press, 1991.

——. "The Ecological Conscience." In *The River of the Mother of God and Other Essays by Aldo Leopold*, edited by Susan L. Flader and J. Baird Callicott, 338–46. Madison: University of Wisconsin Press, 1991.

Li, Ya-Wei. "Recovery." In *Endangered Species Act: Law, Policy, and Perspectives*, edited by Donald C. Baur and Ya-Wei Li, 77–110. Washington, DC: American Bar Association, 2021.

Malcom, Jacob. "Consequences of Resource Limitations on Endangered Species Act Implementation." In *Endangered Species Act: Law, Policy, and Perspectives*, edited by Donald C. Baur and Ya-Wei Li, 417–37. Washington, DC: American Bar Association, 2021.

Malcom, Jacob W., and Ya-Wei Li. "Missing, Delayed, and Old: The Status of ESA Recovery Plans." *Conservation Letters* 11, no. 6 (August 9, 2018). https://doi.org /10.1111/conl.12601.

Male, Timothy. "Making an Asset of Endangered Species Recovery." In *The Codex of the Endangered Species Act, Volume II: The Next Fifty Years*, edited by Lowell E. Baier, John F. Organ, and Christopher E. Segal, 14–29. Lanham, MD: Rowman & Littlefield, 2023.

Marsh, Lindell L., and Robert D. Thornton. "San Bruno Mountain Habitat Conservation Plan." In *Managing Land-Use Conflicts: Case Studies in Special Area Management*, edited by David J. Brower and Daniel S. Carol, 114–39. Durham, NC: Duke University Press, 1987.

Martinez, Michael. "Working Lands for Wildlife: Targeted Landscape-Scale Wildlife Habitat Conservation." *Natural Resources & Environment* 29, no. 3 (2015): 36–39.

Mathiasson, Minna E., and Sandra M. Rehan. "Status Changes in the Wild Bees of North-Eastern North America over 125 Years Revealed through Museum Specimens." *Insect Conservation and Diversity* 12, no. 4 (2019): 278–88. https://doi.org /10.1111/icad.12347.

McGowan, Conor P., Nathan Allan, and David R. Smith. "The Species Status Assessment: A Framework for Assessing Species Status and Risk to Support Endangered Species Management Decisions." In *The Codex of the Endangered Species Act, Volume II: The Next Fifty Years*, edited by Lowell E. Baier, John F. Organ, and Christopher E. Segal, 87–102. Lanham, MD: Rowman & Littlefield, 2023.

Naess, Arne. "The Shallow and the Deep, Long-Range Ecology Movements: A Summary." *Inquiry* 16 (1973): 95–100.

Petersen, Shannon. "Congress and Charismatic Megafauna: A Legislative History of the Endangered Species Act." *Environmental Law* 29 (1999): 463–91.

Rader, Romina, Ignasi Bartomeus, Lucas A. Garibaldi, Michael P. D. Garratt, Brad G. Howlett, Rachael Winfree, Saul A. Cunningham, et al. "Non-Bee Insects Are

Important Contributors to Global Crop Pollination." *PNAS* 113, no. 1 (2016): 146–51. https://doi.org/10.1073/pnas.1517092112.

Rosenberg, Kenneth V., Adriaan M. Dokter, Peter J. Blancher, John R. Sauer, Adam C. Smith, Paul A. Smith, Jessica C. Stanton, et al. "Decline of the North American Avifauna." *Science* 366, no. 6461 (2019): 120–24. https://doi.org/10.1126/science.aaw1313.

Ruckelshaus, Mary, and Donna Darm. "Science and Implementation." In *The Endangered Species Act at Thirty: Conserving Biodiversity in Human-Dominated Landscapes*, vol. 2, edited by J. Michael Scott, Dale D. Goble, and Frank W. Davis, 104–26. Washington, DC: Island Press, 2006.

Ruhl, J. B. "The Battle over Endangered Species Act Methodology." *Environmental Law* 34, no. 2 (2004): 555–603.

———. "While the Cat's Asleep: The Making of the 'New' ESA." *Natural Resources & Environment* 12, no. 3 (1998): 187–90, 224–26.

Smith, Douglas W., Rolf O. Peterson, Daniel R. MacNulty, and Michel Kohl. "The Big Scientific Debate: Trophic Cascades." *Yellowstone Science* 24, no. 1 (June 2016). https://www.nps.gov/articles/the-big-scientific-debate-trophic-cascades.htm.

Stoellinger, Temple, Michael Brennan, Sara Brodnax, Ya-Wei (Jake) Li, Murray Feldman, and Bob Budd. "Improving Cooperative State and Federal Species Conservation Efforts." In *The Codex of the Endangered Species Act, Volume II: The Next Fifty Years*, edited by Lowell E. Baier, John F. Organ, and Christopher E. Segal, 212–37. Lanham, MD: Rowman & Littlefield, 2023.

Stoellinger, Temple, Michael Brennan, Sara Brodnax, Ya-Wei Li, Murray Feldman, and Bob Budd. "Improving Cooperative State and Federal Species Conservation Efforts." *Wyoming Law Review* 20, no. 1 (2020), Article 3.

Thompson, Barton H., Jr. "The Endangered Species Act: A Case Study in Takings & Incentives." *Stanford Law Review* 49, no. 2 (1997): 305–80.

Turner, James Morton. "'The Specter of Environmentalism': Wilderness, Environmental Politics, and the Evolution of the New Right." *Journal of American History* 96, no. 1 (June 2009): 123–48.

Wheeler, Douglas P., and Dale Ratliff. "The Future of Section 10—Multiple Species Habitat Conservation Plans?" In *The Codex of the Endangered Species Act, Volume II: The Next Fifty Years*, edited by Lowell E. Baier, John F. Organ, and Christopher E. Segal, 187–211. Lanham, MD: Rowman & Littlefield, 2023.

Wheeler, Douglas P., and Ryan M. Rowberry. "Habitat Conservation Plans and the Endangered Species Act." In *Endangered Species Act: Law, Policy, and Perspectives*, edited by Donald C. Baur and William Robert Irvin, 220–45. Chicago: American Bar Association, 2010.

Willms, David. "Unlocking the Full Power of Section 4(d) to Facilitate Collaboration and Greater Species Recovery." In *The Codex of the Endangered Species Act, Volume II: The Next Fifty Years*, edited by Lowell E. Baier, John F. Organ, and Christopher E. Segal, 30–50. Lanham, MD: Rowman & Littlefield, 2023.

Wilson, Edward O. "The Current State of Biological Diversity." In *Biodiversity*, edited by Edward O. Wilson, 3–18. Washington, DC: National Academies Press, 1988.

———. "Introduction." In Tony Hiss, *Rescuing the Planet: Protecting Half the Land to Heal the Earth*. New York: Alfred A. Knopf, 2021.

Wilson, Seth M., Elizabeth H. Bradley, and Gregory A. Neudecker. "Learning to Live with Wolves: Community-Based Conservation in the Blackfoot Valley of Montana." *Human–Wildlife Interactions* 11, no. 3 (Winter 2017): 245–57.

Wilson, Seth M., Gregory A. Neudecker, and James J. Jonkel. "Human-Grizzly Bear Coexistance in the Blackfoot River Watershed, Montana: Getting Ahead of the Conflict Curve." In *Large Canivore Conservation: Integrating Science and Policy in the North American West*, edited by Susan G. Clark and Murray B. Rutherford, 177–214. Chicago: University of Chicago Press, 2014.

Wilson, Seth M., and Susan G. Clark. "Resolving Human-Grizzly Bear Conflict: An Integrated Approach in the Common Interest." In *Integrated Resource and Environmental Management: Concepts and Practice*, edited by D. Scott Slocombe and Kevin S. Hanna, 137–63. Oxford: Oxford University Press, 2007.

Wood, Jonathan, and Tate Watkins. "The Future of Habitat: Lessons from the Dusky Gopher Frog Conflict." In *The Codex of the Endangered Species Act, Volume II: The Next Fifty Years*, edited by Lowell E. Baier, John F. Organ, and Christopher E. Segal, 103–24. Lanham, MD: Rowman & Littlefield, 2023.

Zattara, Eduardo E., and Marcelo A. Aizen. "Worldwide Occurrence Records Suggest a Global Decline in Bee Species Richness." *One Earth* 4, no. 1 (2021): 114–23. https://doi.org/10.1016/j.oneear.2020.12.005.

NEWSPAPERS, MAGAZINES, AND ONLINE MEDIA

Arnold, Carrie. "Horseshoe Crab Blood Is Key to Making a Covid-19 Vaccine—but the Ecosystem May Suffer." *National Geographic*, July 2, 2020. https://www.nationalgeographic.com/animals/article/covid-vaccine-needs-horseshoe-crab-blood.

Ashe, Daniel M. "Opinion: Cruelty toward Wolves Is Erasing Conservation Efforts. It's Time to Reinstate Their Protections." *Washington Post*, August 3, 2021. https://www.washingtonpost.com/opinions/2021/08/03/wolves-idaho-montana-cruelty-conservation.

Associated Press. "After Being Released in Idaho, Wolf Is Shot to Death on Ranch." *New York Times*, January 31, 1995, A12.

———. "Wisconsin Opens Early Wolf Hunt after Hunter Group Sued." *Minnesota Public Radio*, February 22, 2021. https://www.mprnews.org/story/2021/02/22/wisconsin-opens-early-wolf-hunt-after-hunter-group-sued.

Barker, Rocky. "Proposed Reform to Endangered Species Act Gets Cool Response." *Idaho Statesman*, August 17, 2008. http://bluefish.org/coolresp.htm.

Becker, Michael. "MSU Presents Presidential Medal to Famed Scientist Edward O. Wilson." *MSU News Service*, April 9, 2009. https://www.montana.edu/news/7071/msu-presents-presidential-medal-to-famed-scientist-edward-o-wilson.

Blevins, Jason. "Proposition 114 Explained: What's at Stake with the Effort to Reintroduce Gray Wolves in Colorado." *Colorado Sun*, September 24, 2020. https://coloradosun.com/2020/09/24/proposition-114-explained-wolf-reintroduction.

Boxall, Betty. "Environmentalists Savor Pombo Defeat as Sign of Power." *Los Angeles Times*, November 9, 2006. https://www.latimes.com/archives/la-xpm-2006-nov-09 -me-defenders9-story.html.

Broder, John M. "Ex-Interior Aide Is Sentenced in Lobbying Case." *New York Times*, June 27, 2007. https://www.nytimes.com/2007/06/27/washington/27griles.html.

Brown, Elizabeth Anne. "Widely Misinterpreted Report Still Shows Catastrophic Animal Decline." *National Geographic*, November 1, 2018. https://www.nationalgeographic .com/animals/article/animal-decline-living-planet-report-conservation-news.

Brown, Susan Jane. "The Return of the Spotted-Owl Wars?" *Seattle Times*, January 22, 2021. https://www.seattletimes.com/opinion/the-return-of-the-spotted-owl-wars.

Chaney, Rob. "Carcass Cleanup Tries to Stay ahead of Grizzlies." *Missoulian*, April 29, 2019. https://missoulian.com/news/local/carcass-cleanup-tries-to-stay-ahead-of -grizzlies/article_7b8cc949-8db0-5afa-98c0-f7fecb9dc32e.html.

Chin, Joshua. "Fighting to Save, and Popularize, San Bruno Mountain." *Bay Nature*, February 26, 2015. https://baynature.org/article/fighting-save-popularize-san-bruno -mountain.

Cohen, Rachel. "25 Years Ago, Wolves Were Reintroduced to Idaho." *Boise State Public Radio*, January 14, 2020. https://www.boisestatepublicradio.org/news/2020-01-14 /25-years-ago-wolves-were-reintroduced-to-idaho.

Daigle, Katy, and Julia Janicki. "Extinction Crisis Puts 1 Million Species on the Brink." *Reuters*, December 23, 2022. https://www.reuters.com/lifestyle/science/extinction -crisis-puts-1-million-species-brink-2022-12-23.

Davis, Tony. "Firebrand Ways: A Visit with One of the Founders of the Center for Biological Diversity." *High Country News*, December 28, 2009. http://www.hcn.org/ issues/41.22/firebrand-ways.

Dellios, Hugh. "Concern at Timber Summit." *Chicago Tribune*, April 2, 1993. https:// www.chicagotribune.com/news/ct-xpm-1993-04-02-9304020374-story.html.

Dialogue. "Revisiting the Tennessee Story of the Snail Darter." Hosted by Chrissy Keuper. Featuring Wayne Starnes, Zygmunt Plater, Peggy Shute, J. R. Shute, and Charles Sims. October 6, 2021. *WUOT*, radio program, 51:04. https://www.wuot.org/news /2021-10-06/dialogue-revisiting-the-tennessee-story-of-the-snail-darter.

Doyle, Michael. "Judge Restores Gray Wolf Protections." *E&E News PM*, February 10, 2022.

———. "Study Brings Owl-Related Job Lossess Down to Earth." *Greenwire*, June 30, 2021. https://www.eenews.net/greenwire/stories/1063736201/print.

"Dramatic Decline in Western Butterfly Populations Linked to Fall Warming." *ScienceDaily*, March 4, 2021. https://www.sciencedaily.com/releases/2021/03 /210304145405.htm.

Duhownik, Joe. "Feds Defend Mexican Wolf Recovery Plan before Ninth Circuit Panel." *Courthouse News Service*, June 5, 2023. https://www.courthousenews.com/feds -defend-mexican-wolf-recovery-plan-before-ninth-circuit-panel.

Dunphy, Siobhán. "Declining Wild Bee Populations Threaten Key Food Crops." *European Scientist*, March 8, 2020. https://www.europeanscientist.com/en/agriculture/ declining-wild-bee-populations-threaten-key-food-crops.

Eilperin, Juliet. "Michelle Obama Makes the Pitch for Pollinators." *Washington Post*, June 3, 2015. https://www.washingtonpost.com/news/post-politics/wp/2015/06/03/michelle-obama-makes-the-pitch-for-pollinators.

Einhorn, Catrin, and Nadja Popovich. "This Map Shows Where Biodiversity Is Most at Risk in America." *New York Times*, March 3, 2022. https://www.nytimes.com/interactive/2022/03/03/climate/biodiversity-map.html.

Elderkin, Susan. "What a Difference a Year Makes." *High Country News*, September 2, 1996. https://www.hcn.org/issues/89/2748.

Farquhar, Brodie. "Pombo Takes on the Endangered Species Act." *High Country News*, October 17, 2005. https://www.hcn.org/issues/308/15840/print_view.

———. "Wolf Reintroduction Changes Ecosystem in Yellowstone." *Outside*, June 30, 2021. https://www.yellowstonepark.com/things-to-do/wildlife/wolf-reintroduction-changes-ecosystem.

Fuller, Jaime. "The Long Fight between the Bundys and the Federal Government, from 1989 to Today." *Washington Post*, January 4, 2016. https://www.washingtonpost.com/news/the-fix/wp/2014/04/15/everything-you-need-to-know-about-the-long-fight-between-cliven-bundy-and-the-federal-government.

Hance, Jeremy. "The Great Insect Dying: The Tropics in Trouble and Some Hope." *Mongabay*, June 10, 2019. https://news.mongabay.com/2019/06/the-great-insect-dying-the-tropics-in-trouble-and-some-hope.

Harney County Sage Grouse CCAA Steering Committee. "2011 Year in Review." *Harney County Sage Grouse CCAA Steering Committee Quarterly Update*, January 2012. https://www.co.harney.or.us/PDF_Files/Sage%20Grouse/Harney%20County%20Sage-grouse%20newsletter%20volume%201%20issue%201%20January%202012.pdf.

Heller, Marc. "Thousands of Tree Species Go Undiscovered, Scientists Say." *Greenwire*, February 3, 2022.

Hiar, Corbin. "Obama Admin Poised to Break Delisting Record." *Greenwire*, May 29, 2015. https://www.eenews.net/stories/1060019305.

Hoffman, David. "Watt Submits Resignation as Interior Secretary." *Washington Post*, October 10, 1983. https://www.washingtonpost.com/archive/politics/1983/10/10/watt-submits-resignation-as-interior-secretary/84ba758c-03f2-439d-8105-0bab802247b9.

Honey, Peter. "Clinton's 'Timber Summit' to Draw Huge Crowds." *Baltimore Sun*, April 2, 1993. https://www.baltimoresun.com/news/bs-xpm-1993-04-02-1993092024-story.html.

Hornblower, Margot. "Carter Signs Bill Forcing Tellico Dam Completion." *Washington Post*, September 26, 1979. https://www.washingtonpost.com/archive/politics/1979/09/26/carter-signs-bill-forcing-tellico-dam-completion/7e57e3c0-d186-4bcf-9930-842c07e21c81.

"How Reintroducing Wolves Helped Save a Famous Park." *BBC Future*. Accessed September 8, 2020. https://www.bbc.com/future/article/20140128-how-wolves-saved-a-famous-park.

Kaeding, Danielle. "Dane County Judge Temporarily Bars Wisconsin's Wolf Hunt, Orders DNR to Set Quota of Zero Wolves." *Wisconsin Public Radio*, October 22,

2021. https://www.wpr.org/dane-county-judge-temporarily-bars-wisconsins-wolf
-hunt-orders-dnr-set-quota-zero-wolves.

Kamen, Al. "Newt Gingrich's Animal Attraction Resurfaces." *Washington Post*, January 27, 2012. https://www.washingtonpost.com/blogs/in-the-loop/post/newt
-gingrichs-animal-attraction-resurfaces/2012/01/26/gIQAofZhTQ_blog.html.

Kenworthy, Tom. "Spotted Owl in the Hands of 'God Squad.'" *Washington Post*, November 20, 1991. https://www.washingtonpost.com/archive/politics/1991/11/20/spotted
-owl-in-the-hands-of-god-squad/097ea868-67b2-44b3-a554-5a17464e1b46.

Kiefer, Michael. "Owl See You in Court." *Phoenix New Times*, August 1, 1996.

Kimbrough, Liz. "Western Monarch Populations Reach Highest Number in Decades." *Mongabay*, January 31, 2023. https://news.mongabay.com/2023/01/monarch
-populations-rebound-but-its-still-a-long-journey-to-recovery.

Lemann, Nicholas. "No People Allowed." *The New Yorker*, November 22, 1999, 96. https://www.newyorker.com/magazine/1999/11/22/no-people-allowed.

Lewis, Sophie. "Animal Populations Worldwide Have Declined Nearly 70% in Just 50 Years, New Report Says." *CBS News*, September 10, 2020. https://www.cbsnews
.com/news/endangered-species-animal-population-decline-world-wildlife-fund
-new-report.

Loomis, Eric. "Think the Spotted Owl Is to Blame for Job Losses? Think Again." *Washington Post*, September 13, 2019. https://www.washingtonpost.com/outlook/2019
/09/13/think-spotted-owl-is-blame-job-losses-think-again.

Lundquist, Laura. "NRCS Designs Grant for Blackfoot Challenge Grizzly Conflict Reduction." *Missoula Current*, November 10, 2021. https://www.kpax.com/
news/western-montana-news/nrcs-designs-grant-for-blackfoot-challenge-grizzly
-conflict-reduction.

Maughan, Ralph. "Defenders Ends Wolf Depredation Payments." *The Wildlife News*, August 31, 2010.

McGivney, Annette. "Moses or Menace?" *Backpacker*, February 2003, 47–53, 90.

"McNerney, Enviros Take Down Richard Pombo." *The Capitol Weekly*, November 9, 2006. https://capitolweekly.net/mcnerney-enviros-take-down-richard-pombo.

McNulty, Timothy J., and Carol Jouzaitis. "Bush, Clinton Try to Balance the Environment and Economy." *Chicago Tribune*, September 15, 1992. https://www
.chicagotribune.com/news/ct-xpm-1992-09-15-9203240170-story.html.

Millsaps, Tommy. "A Look Back: Closing the Tellico Dam Gates." *Monroe County Advocate & Democrat* (Sweetwater, TN), November 30, 2009. https://www.advocateanddemocrat
.com/news/article_1d20abdc-a6e6-5006-9931-389bbe40538e.html.

Mortenson, Eric. "States Adopt Plans as Sage Grouse Listing Decision Approaches." *Capital Press*, April 6, 2015, updated December 13, 2018. https://www.capitalpress
.com/nation_world/nation/states-adopt-plans-as-sage-grouse-listing-decision
-approaches/article_ba299df4-38e5-5cd9-811c-60359aa7947a.html.

National Wildlife Federation. "Conservation News." June 1, 1969.

NWF's Kostyack and NCPPR's Ridenour Go Head to Head over House Species Legislation, aired September 25, 2005, on E&ETV. *E&E News*. https://www.eenews.net/tv/
videos/169/transcript.

Opoien, Jessie. "A Guide to the Legal Battles over Wisconsin's Wolf Hunt." *Capital Times*, September 4, 2021. https://madison.com/ct/news/local/govt-and-politics/election-matters/a-guide-to-the-legal-battles-over-wisconsin-s-wolf-hunt/article_d4b11d0b-d19e-5091-955f-c2a677fedc7c.html.

Organ, John F. "Conservation Without Conflict: Saving Species and Livelihoods." *Fair Chase*, Summer 2020. https://www.boone-crockett.org/conservation-without-conflict-saving-species-and-livelihoods.

Ostlind, Emilene, and Dave Wade. "The Ecology of Fear: Elk Responses to Wolves in Yellowstone Are Not What We Thought." *Western Confluence*, January 6, 2014. https://westernconfluence.org/the-ecology-of-fear-elk-responses-to-wolves-in-yellowstone-are-not-what-we-thought.

Peterson, Christine. "25 Years after Returning to Yellowstone, Wolves Have Helped Stabilize the Ecosystem." *National Geographic*, July 10, 2020. https://www.nationalgeographic.com/animals/article/yellowstone-wolves-reintroduction-helped-stabilize-ecosystem.

Platt, John. "Petition Filed to Protect 404 Southeastern U.S. Species." *Scientific American*, April 22, 2010. https://blogs.scientificamerican.com/extinction-countdown/petition-filed-to-protect-404-southeastern-u-s-species.

"President Signs Bill Reshaping Federal Manpower Programs." *Los Angeles Times*, December 29, 1973.

"President Signs Manpower Bill." *New York Times*, December 29, 1973.

"Proposition 114: The Reintroduction of Gray Wolves, Explained." *Colorado Public Radio*, October 20, 2020. https://www.cpr.org/2020/10/12/vg-2020-colorado-proposition-114-reintroduction-of-gray-wolves-explained.

Quinson, Tim, and Mathieu Benhamou. "Banks Always Backed Fossil Fuel over Green Projects—until This Year." *Bloomberg Green*, May 19, 2021. https://www.bloomberg.com/graphics/2021-wall-street-banks-ranked-green-projects-fossil-fuels.

Reuters. "Grizzlies in Yellowstone Area to Leave Endangered Species List." *Washington Post*, March 23, 2007, A8.

Revkin, Andrew. "E.O. Wilson at 90: The Conservation Legend Shares Dreams for the Future." *National Geographic*, June 10, 2019. https://www.nationalgeographic.com/environment/article/eo-wilson-conservation-legend-90-save-space-for-nature-save-planet.

Ridler, Keith. "Legislation Aims to Kill 90% of Wolves Roaming Idaho." *Associated Press*, April 20, 2021. https://apnews.com/article/wildlife-legislation-environment-wolves-bills-ccb1f49f74f0781ebbacd10c48f5abc1.

Rinde, Meir. "Richard Nixon and the Rise of American Environmentalism." *Distillations* (blog), *Science History Institute*, June 2, 2017. https://www.sciencehistory.org/distillations/richard-nixon-and-the-rise-of-american-environmentalism.

Rothenberg, Stuart. "Utah's Frank Moss Was a Symbol of Nation's Realignment." *Roll Call*, February 5, 2003. https://rollcall.com/2003/02/05/utahs-frank-moss-was-a-symbol-of-nations-realignment.

Rott, Nathan. "The World Lost Two-Thirds of Its Wildlife in 50 Years. We Are to Blame." *NPR*, September 10, 2020. https://www.npr.org/2020/09/10/911500907/the-world-lost-two-thirds-of-its-wildlife-in-50-years-we-are-to-blame.

Schiller, Ben. "Investors Are Pouring Money into Conservation Efforts (It's Actually a Great Investment)." *Fast Company*, February 2, 2017. https://www.fastcompany.com/3067575/investors-are-pouring-money-into-conservation-efforts-its-actually-a-.

Scott, Austin. "Nixon Signs Bill to Give States Manpower Funds." *Washington Post*, December 29, 1973.

Sidder, Aaron. "New Map Highlights Bee Population Declines across the U.S." *Smithsonian*, February 23, 2017. https://www.smithsonianmag.com/smart-news/new-map-highlights-bee-population-declines-across-us-180962268.

Somashekhar, Sandhya. "Gingrich Wild about Zoos." *New York Times*, December 9, 2011. https://www.washingtonpost.com/politics/2011/12/08/gIQAVb1yiO_story.html.

Steinhauer, Nathalie. "United States Honey Bee Colony Losses 2022–23: Preliminary Results from the Bee Informed Partnership." *Bee Informed*, June 22, 2023. https://beeinformed.org/2023/06/22/united-states-honey-bee-colony-losses-2022-23-preliminary-results-from-the-bee-informed-partnership.

Taylor, Christie. "Wisconsin Wolf Hunt Overshoots Quota, Worrying Conservationists." *Madison Public Radio*, March 5, 2021.

Timber Wars. Podcast audio. *National Public Radio*. https://www.npr.org/podcasts/906829608/timber-wars.

University of New Hampshire. "More Than Dozen Wild Bee Species Declining in Northeastern U.S." *ScienceDaily*, April 9, 2019. https://www.sciencedaily.com/releases/2019/04/190409093801.htm.

Vigdor, Neil. "Wisconsin More Than Doubles Wolf-Hunting Quota, Angering Conservationists." *New York Times*, August 11, 2021. https://www.nytimes.com/2021/08/11/us/wisconsin-wolves-hunting-kill.html.

Warren, Lydia. "Cages Built to Protect Kids from Wolves at New Mexico Bus Stops under Fire for 'Demonizing' the Endangered Animals." *Daily Mail*, October 30, 2013. https://www.dailymail.co.uk/news/article-2479999/Cages-built-protect-kids-wolves-New-Mexico-bus-stops-demonizing-endangered-animals.html.

Weber, Wendi. "ESA Helps Write New Chapter for Peter Cottontail." *Huffington Post*, September 22, 2015. https://newenglandcottontail.org/news/esa-helps-write-new-chapter-peter-cottontail.

Welsch, Jeff. "Keeping Montana . . . Montana." *Montana Quarterly*, Fall 2013, 8–16.

Yakas, Ben, "Wyoming Wants to Bring Wolves to Central Park." *Gothamist* (blog), *New York Public Radio*, February 12, 2012. https://gothamist.com/news/wyoming-wants-to-bring-wolves-to-central-park.

Yong, Ed. "Wait, Have We Really Wiped Out 60 Percent of Animals?" *The Atlantic*, October 31, 2018. https://www.theatlantic.com/science/archive/2018/10/have-we-really-killed-60-percent-animals-1970/574549.

LEGISLATIVE MATERIALS

Balanced Economic and Environmental Priorities Act of 1991. H.R. 4058. 102nd Cong. 1991.

Barriers to Endangered Species Act Delistings, Part II: Joint Hearing before the Subcommittee on the Interior and the Subcommittee on Health Care, Benefits and Administrative Rules of the House Committee on Oversight and Government Reform. 114th Cong. 2016.

Congressional Record.

Defining Species Conservation Success: Tribal, State and Local Stewardship vs. Federal Courtroom Battles and Sue-and-Settle Practices: Oversight Hearing before the House Committee on Natural Resources. 113th Cong. 2013.

Department of the Interior and Related Agencies Appropriations Act, 2002. Pub. L. No. 107-63, 115 Stat. 414. 2001.

Emergency Supplemental Appropriations and Rescissions for the Department of Defense to Preserve and Enhance Military Readiness Act of 1995. Pub. L. No. 104-6, 109 Stat. 73. 1995.

Endangered Species Act of 1973. H.R. Rep. No. 93-740. 1973 (Conf. Rep.).

Endangered Species Act of 1973, Pub. L. No. 93-205, 87 Stat. 884. 1973.

Endangered Species Act Amendments of 1978. Pub. L. No. 95-632, 92 Stat. 3751. 1978.

Endangered Species Act Amendments of 1982. H.R. Rep. No. 97-835. 1982 (Conf. Rep.).

Endangered Species Act Amendments of 1988. H.R. Rep. No. 100-928. 1988 (Conf. Rep.).

The Endangered Species Act: How Litigation Is Costing Jobs and Impeding True Recovery Efforts: Oversight Hearing before the House Committee on Natural Resources. 112th Cong. 2011.

Endangered Species Act Oversight: Hearings before the Subcommittee on Resource Protection of the Senate Committee on Environment and Public Works. 95th Cong. 1977.

Endangered Species Act Procedural Reform Amendments of 1993. S. 1521. 103rd Cong. 1993.

Endangered Species Act: Washington, DC—Part III: Hearing on the Impact of the Endangered Species Act on the Nation before the Task Force on the Endangered Species Act of the House Committee on Resources. 104th Cong. 1995.

Endangered Species: Hearings on H.R. 37, H.R. 470, H.R. 471, H.R. 1461, H.R. 1511, H.R. 2669, H.R. 2735, H.R. 3310, H.R. 3696, H.R. 3795, H.R. 4755, H.R. 2169, and H.R. 4758 before the Subcommittee on Fisheries and Wildlife Conservation and the Environment of the House Committee on Merchant Marine and Fisheries. 93rd Cong. 1973.

Endangered Species Recovery Act of 1997. S. 1180. 105th Cong. 1997.

Energy and Water Development Appropriation Act, 1980. Pub. L. No. 96-69, 93 Stat. 437. 1979.

ESA Decisions by Closed-Door Settlement: Short-Changing Science, Transparency, Private Property, and State & Local Economies: Legislative Hearing on H.R. 4315, H.R. 4316, H.R. 4317, and H.R. 4318 before the House Committee on Natural Resources. 113th Cong. 2013.

Fish and Wildlife Miscellaneous, Part 1: Hearings on H.R. 2817, H.R. 4141, H.R. 4741, H.R. 3633, H.R. 8394, H.R. 7749, and H.R. 4658 before the Subcommittee on Fisheries and Wildlife Conservation and the Environment of the House Committee on Merchant Marine and Fisheries. 95th Cong. 1977.

House Committee on Merchant Marine and Fisheries. Endangered and Threatened Species Conservation Act of 1973. H.R. Rep. No. 93-412. 1973.

Human Protection Act of 1991. H.R. 3092. 102nd Cong. 1991.

Partners for Fish and Wildlife Act. Pub. L. No. 109-294, 120 Stat. 1351. 2006.

Stenographic Transcript of Hearings before the Committee on Merchant Marine and Fisheries, House of Representatives, 95th Congress, Markup Session on H.R. 13807 and H.R. 14104, September 19, 1978.

Threatened and Endangered Species Recovery Act of 2005. H.R. 3824. 109th Cong. 2005.

William M. (Mac) Thornberry National Defense Authorization Act for Fiscal Year 2021. Pub. L. No. 116-283, 134 Stat. 3388. 2021.

ADMINISTRATIVE, REGULATORY, AND OTHER EXECUTIVE MATERIALS

12-Month Finding for the Monarch Butterfly. 85 Fed. Reg. 81813. December 17, 2020.

12-Month Finding on a Petition to List Greater Sage-Grouse (*Centrocercus urophasianus*) as an Endangered or Threatened Species. 80 Fed. Reg. 59858. October 2, 2015.

12-Month Finding on a Petition to List the New England Cottontail as an Endangered or Threatened Species. 80 Fed. Reg. 55286. September 15, 2015.

90-Day Finding for Two Petitions to List the Gray Wolf in the Western United States. 86 Fed. Reg. 51857. September 17, 2021.

90-Day Petition Finding and Initiation of Status Review, Northern Spotted Owl. 52 Fed. Reg. 34396. September 11, 1987.

Announcement of Final Policy for Candidate Conservation Agreements with Assurances. 64 Fed. Reg. 32726. June 17, 1999.

Announcement of Final Safe Harbor Policy. 64 Fed. Reg. 32717. June 17, 1999.

Bureau of Land Management. Instruction Memorandum 2018-093: Compensatory Mitigation. July 24, 2018.

Buttolph, Lita P., William Kay, Susan Charnley, Cassandra Moseley, and Ellen M. Donoghue. *Northwest Forest Plan—The First 10 Years (1994–2003): Socioeconomic Monitoring of the Olympic National Forest and Three Local Communities.* General Technical Report PNW-GTR-679. U.S. Forest Service, July 2006. https://www.fs.fed.us/pnw/pubs/pnw_gtr679.pdf.

Compensatory Mitigation for Losses of Aquatic Resources. 73 Fed. Reg. 19594. April 10, 2008.

Compensatory Mitigation Mechanisms. 87 Fed. Reg. 45076. July 27, 2022.

Determination of Critical Habitat for the Northern Spotted Owl. 57 Fed. Reg. 1796. January 15, 1992.

Determination of Threatened Status for the Northern Spotted Owl. 55 Fed. Reg. 26114. June 26, 1990.

Endangered Species Act Compensatory Mitigation Policy. 81 Fed. Reg. 95316. December 27, 2016.

Endangered Species Act Compensatory Mitigation Policy. 83 Fed. Reg. 36469. July 30, 2018.

Endangered Species Permit: Notice of Receipt of Application. 39 Fed. Reg. 18483. May 28, 1974.

Endangered Species Permit: Receipt of Addendum to Application. 41 Fed. Reg. 8515. February 27, 1976.

Endangered Species Permits: Official Action. 41 Fed. Reg. 9576. March 5, 1976.

Endangered Species Status for Rusty Patched Bumble Bee. 82 Fed. Reg. 3186. January 11, 2017.

Enhancement of Survival and Incidental Take Permits. 88 Fed. Reg. 8380. February 9, 2023.

Establishment of a Nonessential Experimental Population of the Mexican Gray Wolf in Arizona and New Mexico. 63 Fed. Reg. 1752. January 12, 1998.

Exec. Order No. 13,783: Promoting Energy Independence and Economic Growth. 82 Fed. Reg. 16093. March 28, 2017.

Final Rule to Remove the American Peregrine Falcon from the Federal List of Endangered and Threatened Wildlife, and to Remove the Similarity-of-Appearance Provision for Free-Flying Peregrines in the Conterminous United States. 64 Fed. Reg. 46542. August 25, 1999.

Finding on Northern Spotted Owl Petition. 52 Fed. Reg. 48552. December 23, 1987.

Greater Sage-Grouse Programmatic Candidate Conservation Agreement with Assurances for Private Rangelands in Harney County, Oregon between the Harney Soil and Water Conservation District and the United States Fish and Wildlife Service. April 25, 2014. https://www.fws.gov/oregonfwo/toolsforlandowners/ habitatconservationplans/Documents/5.13.14%20Harney%20SWCD%20CCAA -Final.pdf.

National Park Service. *Yellowstone Wolf Project Biennial Report: 1995 and 1996.* https:// home.nps.gov/yell/learn/nature/upload/wolfrep95-96.pdf.

Natural Resources Conservation Service. *Working Lands for Wildlife Obligations, All Years All Species.* 2019.

Possible Effects of Court Decision on Grizzly Bear Recovery in the Conterminous United States. 82 Fed. Reg. 57698. December 7, 2017.

Post-Delisting Monitoring Results for the American Peregrine Falcon (Falco peregrinus anatum), 2003. 71 Fed. Reg. 60563. October 13, 2016.

Proposed Determination of Critical Habitat for the Mexican Spotted Owl. 69 Fed. Reg. 63162. December 7, 1994.

Reinstatement of Protections for the Gray Wolf in the Western Great Lakes in Compliance with Settlement Agreement and Court Order. 74 Fed. Reg. 47483. September 16, 2009.

Removal of the Louisiana Black Bear from the Federal List of Endangered and Threatened Wildlife and Removal of Similarity-of-Appearance Protections for the American Black Bear. 81 Fed. Reg. 13124. March 11, 2016.

Removing the Gray Wolf (*Canis lupus*) from the List of Endangered and Threatened Wildlife. 85 Fed. Reg. 69778. November 3, 2020.

Removing the Greater Yellowstone Ecosystem Population of Grizzly Bears from the Federal List of Endangered and Threatened Wildlife. 82 Fed. Reg. 30502. June 30, 2017.

Removing the Kirtland's Warbler from the Federal List of Endangered and Threatened Wildlife. 84 Fed. Reg. 54436. October 9, 2019.

Removing the Yellowstone Distinct Population Segment of Grizzly Bears from the Federal List of Endangered and Threatened Wildlife; 90-Day Finding on a Petition to List as Endangered the Yellowstone Distinct Population Segment of Grizzly Bears. 72 Fed. Reg. 14866. March 29, 2007.

Review of 2017 Final Rule, Greater Yellowstone Ecosystem Grizzly Bears. 83 Fed. Reg. 18737. April 30, 2018.

Revision of Regulations for Interagency Cooperation. 88 Fed. Reg. 40753. June 22, 2023.

Revision to the Nonessential Experimental Population of the Mexican Wolf. 86 Fed. Reg. 59953. October 29, 2021.

Revision to the Regulations for the Nonessential Experimental Population of the Mexican Wolf; Final Rule. 80 Fed. Reg. 2512. January 16, 2015.

Revisions to the Regulations for Petitions. 81 Fed. Reg. 66462. September 27, 2016.

Thomas, Jack Ward, Eric D. Forsman, Joseph B. Lint, E. Charles Meslow, Barry B. Noon, and Jared Verner. *A Conservation Strategy for the Northern Spotted Owl: Report of the Interagency Scientific Committee to Address the Conservation of the Northern Spotted Owl.* 1990.

Threatened Species Status with Section 4(d) Rule for the Upper Coosa River Distinct Population Segment of Frecklebelly Madtom and Designation of Critical Habitat. 85 Fed. Reg. 74050. November 19, 2020.

Threatened Species Status with Section 4(d) Rule for the Upper Coosa River Distinct Population Segment of Frecklebelly Madtom and Designation of Critical Habitat. 88 Fed. Reg. 13038. March 2, 2023.

U.S. Department of the Interior. *Budget Justifications and Performance Information Fiscal Year 2021: Fish and Wildlife Service.* 2021. https://www.fws.gov/budget/2021/FY2021-FWS-Budget-Justification.pdf.

———. *Secretarial Order No. 3349: American Energy Independence.* March 29, 2017.

———. *Secretarial Order No. 3360: Rescinding Authorities Inconsistent with Secretary's Order 3349, "American Energy Independence."* December 22, 2017.

U.S. Department of the Interior, U.S. Department of Agriculture, U.S. Department of Commerce, and Council on Environmental Quality. *Conserving and Restoring America the Beautiful.* 2021. https://www.doi.gov/sites/doi.gov/files/report-conserving-and-restoring-america-the-beautiful-2021.pdf.

U.S. Department of the Interior Office of Inspector General. "Report of Investigation: Julie MacDonald, Deputy Assistant Secretary, Fish, Wildlife and Parks." Accessed September 22, 2023. https://grist.org/wp-content/uploads/2007/03/doi-ig-report_jm.pdf.

U.S. Fish and Wildlife Service. *2022 National Survey of Fishing, Hunting, and Wildlife–Associated Recreation.* September 2023. https://digitalmedia.fws.gov/digital/api/collection/document/id/2321/download.

———. *Draft Candidate Conservation Agreements with Assurances Handbook.* June 2003.

———. *Guidance for the Establishment, Use, and Operation of Conservation Banks.* May 2, 2003.

———. *A Habitat Conservation Plan to Encourage the Voluntary Restoration and Enhancement of Habitat for the Red–Cockaded Woodpecker on Private and Certain Other Land in the Sandhills Region of North Carolina by Providing "Safe Harbor" to Participating Landowners.* 1995. https://ecos.fws.gov/docs/plan_documents/tsha/tsha_451.pdf.

———. *Mexican Wolf Recovery Program: Progress Report #25.* 2023. https://www.fws.gov/sites/default/files/documents/Mexican-Wolf-2022-Progress-Report-Final.pdf.

———. *Nationwide Candidate Conservation Agreement for Monarch Butterfly on Energy and Transportation Lands: An Integrated Candidate Conservation Agreement with Assurances (CCAA) and Candidate Conservation Agreement (CCA).* March 2020. https://www.fws.gov/savethemonarch/pdfs/Final_CCAA_040720_Fully%20Executed.pdf.

———. *Protecting America's Living Heritage: A Fair, Cooperative and Scientifically Sound Approach to Improving the Endangered Species Act.* March 6, 1995. http://www.fws.gov/policy/npi96_06.pdf.

———. *Recovery Outline for the Frecklebelly Madtom.* May 25, 2023. https://ecos.fws.gov/docs/recovery_plan/20230524_Frecklebelly%20madtom%20Recovery%20Outline.pdf.

———. *Snail Darter Recovery Plan, Second Revision.* December 12, 1982.

U.S. Fish and Wildlife Service and National Marine Fisheries Service. *Habitat Conservation Planning and Incidental Take Permit Processing Handbook.* 2016.

U.S. Fish and Wildlife Service Mitigation Policy. 83 Fed. Reg. 36472. July 30, 2018.

U.S. Fish and Wildlife Service Mitigation Policy and Endangered Species Act Compensatory Mitigation Policy. 88 Fed. Reg. 31000. May 15, 2023.

U.S. Forest Service. *Greater Sage-Grouse Proposed Land Management Plan Amendments (LMPA) and Draft Environmental Impact Statement (DEIS) for the Intermountain and Rocky Mountain Regions.* October 2018. https://www.fs.usda.gov/Internet/FSE_DOCUMENTS/fseprd598215.pdf.

U.S. Government Accountability Office. *Wildlife and Sport Fish Restoration: Competitive Grant Programs Managed Consistently with Relevant Regulations, but Monitoring Could Be Improved.* GAO-18-303. February 2018. https://www.gao.gov/assets/gao-18-303.pdf.

COURT CASES

Alabama–Tombigbee River Coalition v. Kempthorne, 477 F.3d 1250 (11th Cir. 2007).

Center for Biological Diversity v. Jewell, 2018 WL 1586651 (D. Ariz. 2018).

Crow Indian Tribe v. United States, 343 F. Supp. 3d 999 (D. Mont 2018).

Crow Indian Tribe v. United States, 965 F.3d 662 (9th Cir. 2020).

Defenders of Wildlife v. Hall, 565 F. Supp. 2d 1160 (D. Mont. 2008).

Defenders of Wildlife v. Salazar, 729 F. Supp. 2d 1207 (D. Mont. 2010).

Defenders of Wildlife v. Secretary, Department of the Interior, 354 F. Supp. 2d 1156 (D. Ore 2005).

Environmental Defense Fund v. Tennessee Valley Authority, 339 F. Supp. 806 (E.D. Tenn. 1972).

Environmental Defense Fund v. Tennessee Valley Authority, 371 F. Supp. 1004 (E.D. Tenn. 1973).

Greater Yellowstone Coalition, Inc. v. Servheen, 672 F. Supp. 2d 1105 (D. Mont. 2009).

Greater Yellowstone Coalition, Inc. v. Servheen, 665 F.3d 1015 (9th Cir. 2011).

Hill v. Tennessee Valley Authority, 419 F. Supp. 753 (E.D. Tenn. 1976).

Hill v. Tennessee Valley Authority, 549 F.2d 1064 (6th Cir. 1977).

Humane Society of the United States v. Jewell, 76 F. Supp. 3d 69 (D. D.C.2014).

Humane Society of the United States v. Kempthorne, 579 F. Supp. 2d 7 (D. D.C. 2008).

Humane Society of the United States v. Zinke, 865 F.3d 585 (D.C. Cir. 2017).

In re: Endangered Species Act § 4 Deadline Litigation—MDL No. 2165, No. 1:2010-mc-00377-EGS (D.D.C. 2010).

National Association of Home Builders v. Babbitt, 130 F.3d 1041 (D.C. Cir. 1997).

National Wildlife Federation v. Norton, 386 F. Supp.2d 553 (D. Vt. 2005).

Northern Spotted Owl v. Hodel, 716 F. Supp. 479 (W.D. Wash., 1988).

Tennessee Valley Authority v. Hill, 437 U.S. 153 (1978).

Weyerhaeuser Co. v. U.S. Fish & Wildlife Service, 139 S. Ct. 361 (2018).

INTERVIEWS AND PERSONAL COMMUNICATIONS

Allison, Lesli. Interview by Christopher E. Segal. February 17, 2017.

Ashe, Daniel. Interview by Lowell E. Baier and Christopher E. Segal. December 20, 2016.

Barry, Donald J. Interviews by Lowell E. Baier and Christopher E. Segal. January 25, 2019; February 27, 2019; and November 2, 2021.

Beard, Bruce. Interview by Christopher E. Segal. July 10, 2019.

Bohlen, E. U. Curtis "Buff." Interviews by Lowell E. Baier and Christopher E. Segal. August 7, 2019, and December 12, 2020.

Bullock, Jimmy. Interview by Christopher E. Segal. January 26, 2021.

Cummins, James L. Interview by Christopher E. Segal. March 6, 2019.

Dingell, John D., Jr. Interview by Lowell E. Baier. April, 29, 2016.

Frazer, Gary. E-mail to Lowell E. Baier. June 23, 2021.

Gober, Donald "Pete." Interview by Lowell E. Baier and Christopher E. Segal. May 25, 2018.

Jackson, Wendy. Interview by Christopher E. Segal. January 28, 2019.

Lyons, James R. Interview by Lowell E. Baier and Christopher E. Segal. October 9, 2018.

Male, Tim. Interview by Lowell E. Baier and Christopher E. Segal. March 5, 2020.

Marsh, Lindell L. Interview by Christopher E. Segal. November 30, 2018.

Mayer, Kenneth E. Interview by Christopher E. Segal. March 24, 2014.

Perrotti, Louis. Interview by Christopher E. Segal. January 29, 2021.

Potter, Frank. Interviews by Lowell E. Baier. November 20 and 23, 2021.

Reed, Nathaniel P. Interview by Lowell E. Baier and Christopher E. Segal. May 18, 2016.

Stiver, San. Interview by Christopher E. Segal. March 25, 2014.

Talbot, Lee M. Interview by Lowell E. Baier and Christopher E. Segal. March 29, 2016.

Talbott, Scott. Interview by Lowell E. Baier and Christopher E. Segal. July 9, 2017.

Tur, Anthony. E-mail to Christopher E. Segal. February 17, 2021.

Walker, Wayne. Interview by Christopher E. Segal. October 5, 2018.

Weber, Wendi. Interview by Christopher E. Segal and Lowell E. Baier. January 21, 2021.

Wheeler, Douglas P. Interviews by Lowell E. Baier and Christopher E. Segal. March 24, 2017, and November 23, 2021.

PRESENTATIONS AND CONFERENCE PAPERS

Ashe, Daniel M. "Keynote Address." Presentation at the 81st North American Wildlife and Natural Resources Conference (Pittsburgh, PA). March 15, 2016.

Hamilton, Healy. "The Role of Biodiversity in a Resilient America." Presentation at Wildlife Corridors and Saving America's Biodiversity with E. O. Wilson (Washington, DC). October 24, 2017.

Phifer, Paul. Presentation at the Endangered Species Act Luncheon Roundtable (Washington, DC). March 5, 2020.

Stiver, San. "Background on the Executive Oversight Committee." Presentation to the Sagebrush Executive Oversight Committee, Association of Fish and Wildlife Agencies Annual Meeting (St. Paul, MN). September 24, 2019.

PRESS RELEASES

Center for Biological Diversity. "Conservation Groups Challenge Kill-at-Will Policy for Wyoming Wolves." Press release, September 10, 2012. https://www .biologicaldiversity.org/news/press_releases/2012/wolves-09-10-2012.html.

———. "Historic Accomplishment: Snail Darter Recovered." Press release, August 31, 2021. https://biologicaldiversity.org/w/news/press-releases/historic -accomplishment-snail-darter-recovered-2021-08-31.

Defenders of Wildlife. "Defenders Shifts Focus to Wolf Coexistence Partnerships." Press release, August 20, 2010. https://defenders.org/newsroom/defenders-shifts-focus -wolf-coexistence-partnerships.

Ecosystem Marketplace. "Private Investment in Conservation Reaches $8.2 Billion." Press release, January 11, 2017. https://www.ecosystemmarketplace.com/articles/ private-investment-in-conservation-reaches-8-2-billion.

NatureServe. "Over One-Third of Biodiversity in the United States Is at Risk of Disappearing." Press release, February 6, 2023. https://www.natureserve.org/news -releases/over-one-third-biodiversity-united-states-risk-disappearing.

ONLINE SOURCES

Alaska Department of Fish and Game. "Wolf (Canis lupus) Species Profile." Accessed November 3, 2021. https://www.adfg.alaska.gov/index.cfm?adfg=wolf.main.

Bansal, Tima, and Devika Agarwal. "Corporate Sustainability—Meaning, Examples and Importance." *Network for Business Sustainability*. March 10, 2021. https://www.nbs .net/articles/corporate-sustainability-meaning-examples-and-importance.

Ben & Jerry's. "How Ben & Jerry's Is Fighting Global Warming." February 24, 2015. https://www.benjerry.com/values/issues-we-care-about/climate-justice/ fighting-global-warming.

Blackfoot Challenge. "History." Accessed December 15, 2020. https://blackfootchallenge .org/history.

Center for Biological Diversity. "America's Gray Wolves: A Long Road to Recovery." Accessed September 8, 2020. https://www.biologicaldiversity.org/campaigns/gray _wolves/index.html.

Colbert, Angela. "A Global Biodiversity Crisis: How NASA Satellites Help Track Changes to Life on Earth." *National Aeronautics and Space Administration*. May 22, 2023. https://climate.nasa.gov/news/3265/a-global-biodiversity-crisis-how-nasa -satellites-help-track-changes-to-life-on-earth.

Colorado Parks and Wildlife. "Gray and Mexican Wolves." June 21, 2023. https://cpw .state.co.us/learn/Pages/SOC-Wolves.aspx.

———. "Wolves in Colorado FAQ." Accessed June 21, 2023. https://cpw.state.co.us/ learn/Pages/Wolves-in-Colorado-FAQ.aspx.

Defenders of Wildlife. "Gray Wolf." Accessed September 8, 2020. https://defenders.org /wildlife/gray-wolf.

Endangered Species Coalition. "Lee M. Talbot, a Personal Perspective on the Endangered Species Act of 1973 (ESA) and the Convention on International Trade in Endangered Species of Wild Fauna and Flora (CITES)." Accessed July 28, 2018. http://www.endangered.org/campaigns/wild-success-endangered-species -act-at-40/lee-talbot.

Environmental Policy Innovation Center. "A Guide to the Revised Endangered Species Regulations." Accessed July 6, 2021. http://policyinnovation.org/esaregs19.

Environmental Protection Agency. "Federal Pollinator Health Task Force: EPA's Role." Updated January 19, 2017. https://19january2017snapshot.epa.gov/pollinator -protection/federal-pollinator-health-task-force-epas-role_.html.

———. "Future of Climate Change." January 19, 2017. https://19january2017snapshot .epa.gov/climate-change-science/future-climate-change_.html.

Fitzgerald, Patrick. *Monarch Conservation in America's Cities*. National Wildlife Federation. 2015. http://www.nwf.org/~/media/PDFs/Garden-for-Wildlife/Monarch -Conservation-in-Americas-Cities_Guide-121715.pdf.

Forest2Market. "New Report Details the Economic Impact of US Forest Products Industry." May 8, 2019. https://www.forest2market.com/blog/new-report-details -the-economic-impact-of-us-forest-products-industry.

Future Learn. "What Is Corporate Sustainability and Why Is It Important?" September 10, 2021. https://www.futurelearn.com/info/blog/what-is-corporate-sustainability.

Giaimo, Cara. "Yellowstone Owes Its Early Success to Public Bear Feeding." *Atlas Obscura*. August 31, 2016. https://www.atlasobscura.com/articles/yellowstone-owes -its-early-success-to-public-bear-feeding.

Habitat Conservation Assistance Network. "Monarch Butterfly." Accessed March 8, 2022. https://www.habitatcan.org/Monarch-Butterfly.

Half-Earth Project. "Discover Half-Earth." Accessed October 28, 2021. https://www.half-earthproject.org/discover-half-earth.

International Union for Conservation of Nature. "IUCN Red List of Threatened Species." Accessed December 17, 2023. https://www.iucnredlist.org/search.

International Wolf Center. "Are Wolves Dangerous to Humans?" Accessed June 20, 2023. https://wolf.org/wolf-info/factsvsfiction/are-wolves-dangerous-to-humans.

IUCN Red List. "Brown Bear (*Ursus arctos*)." June 21, 2023. https://www.iucnredlist.org/species/pdf/121229971/attachment.

———. "Grey Wolf (*Canis lupus*)." March 9, 2022. https://www.iucnredlist.org/species/pdf/133234888/attachment.

"Journals of the Lewis & Clark Expedition, March 2, 1806." Accessed August 11, 2017. https://lewisandclarkjournals.unl.edu/item/lc.jrn.1806-03-02#n28030201.

Kopec, Kelsey, and Lori Ann Burd. *Pollinators in Peril. Center for Biological Diversity.* February 2017. https://www.biologicaldiversity.org/campaigns/native_pollinators/pdfs/Pollinators_in_Peril.pdf.

Living with Wolves. "Are Wolves Killing Lots of Cattle and Sheep?" Accessed September 8, 2020. https://www.livingwithwolves.org/portfolio/are-wolves-killing-lots-of-cattle-and-sheep.

Lobos of the Southwest. "In the News: Revised Rules Let Mexican Wolves Roam Expanded Territory." Accessed April 28, 2021. https://mexicanwolves.org/in-the-news-revised-rules-let-mexican-wolves-roam-expanded-territory.

MacNeil, Caeleigh. "How Wolves Saved the Foxes, Mice and Rivers of Yellowstone National Park." *Earthjustice.* October 26, 2016. https://earthjustice.org/blog/2015-july/how-wolves-saved-the-foxes-mice-and-rivers-of-yellowstone-national-park.

Make Way for Monarchs. "Michelle Obama Creates First-Ever 'Pollinator Garden' at the White House." April 30, 2014. https://makewayformonarchs.org/2014/04/michelle-obama-creates-first-ever-pollinator-garden-at-the-white-house.

Male, Timothy. "Saving the Monarch Butterfly." *Sand County Foundation.* April 9, 2020. https://sandcountyfoundation.org/news/2020/saving-the-monarch-butterfly.

Midwest Association of Fish and Wildlife Agencies. "Q&A Messages for Mid-America Monarch Conservation Strategy." Accessed July 19, 2023. http://www.mafwa.org/?page_id=2414.

National Animal Interest Alliance. "Quotes from the Leaders of the Animal Rights Movement." Accessed August 8, 2010. http://www.naiaonline.org/body/article/archives/animalrightsquote.htm.

National Audubon Society. "12 Ways the Inflation Reduction Act Will Benefit Birds and People." August 17, 2022. https://www.audubon.org/news/12-ways-inflation-reduction-act-will-benefit-birds-and-people.

National Ocean Service. "2022 Sea Level Rise Technical Report." Accessed July 20, 2023. https://oceanservice.noaa.gov/hazards/sealevelrise/sealevelrise-tech-report.html.

National Park Service. "Yellowstone: Grizzly Bears & the Endangered Species Act." Updated March 2, 2020. https://www.nps.gov/yell/learn/nature/bearesa.htm.

———. "Yellowstone: Wolf Restoration." Updated May 21, 2020. https://www.nps.gov/yell/learn/nature/wolf-restoration.htm.

National Wildlife Federation. "Mayors' Monarch Pledge: Program Overview." Accessed July 18, 2023. https://www.nwf.org/MayorsMonarchPledge/About/Overview.

———. "Saving the Greater Sage-Grouse." Accessed May 7, 2018. https://www.nwf.org/Our-Work/Wildlife-Conservation/Greater-Sage-Grouse.

Natural Resources Conservation Servce. "Monarch Butterflies." Accessed July 19, 2023. https://www.nrcs.usda.gov/programs-initiatives/monarch-butterflies.

———. "Working Lands for Wildlife." Accessed June 5, 2023. https://www.wlfw.org.

Nature Needs Half. "History." Accessed October 28, 2021. https://natureneedshalf.org/who-we-are/history.

Nowierski, Robert M. "Pollinators at a Crossroads." *U.S. Department of Agriculture*. July 29, 2021. https://www.usda.gov/media/blog/2020/06/24/pollinators-crossroads.

Patagonia. "Environmental Activism." Accessed November 19, 2021. https://www.patagonia.com/activism.

"Politics of Extinction, Attacks on the Endangered Species Act." Accessed July 15, 2021. http://www.biologicaldiversity.org/compaigns/esa_attacks/table.html.

Public Lands Foundation. "Landscape Stewardship Awards—2014." September 9, 2014. https://publicland.org/awards/harney-county-soil-and-water-conservation-district.

"Resource Management Service, LLC." Accessed November 19, 2021. https://resourcemgt.com.

Roger Williams Park Zoo. "New England Cottontail." Accessed January 28, 2021. https://www.rwpzoo.org/conservation-local/new-england-cottontail.

South Carolina Forestry Commission. "Forest Management Facts." Accessed November 16, 2021. https://www.state.sc.us/forest/refmgt.htm.

Southeast Conservation Adaptation Strategy. "About SECAS." Accessed February 23, 2021. https://secassoutheast.org/about.

Stuebner, Steve. "Wolves Part 4: Unforeseen Impacts Caused by Wolves in Idaho." *Life on the Range*. Accessed June 21, 2022. https://idrange.org/range-stories/north-central-idaho/unforeseen-impacts-caused-by-wolves-in-idaho.

The Nature Conservancy. "Companies Investing in Nature." Accessed November 19, 2021. https://www.nature.org/en-us/about-us/who-we-are/how-we-work/working-with-companies/companies-investing-in-nature1.

———. "Water Funds Toolbox." Accessed February 26, 2021. https://waterfundstoolbox.org.

"UIC Leads Largest Nationwide Effort to Protect the Monarch Butterfly." *UIC Today*. April 8, 2020. https://today.uic.edu/uic-leads-largest-nationwide-effort-to-protect-the-monarch-butterfly.

United States Environmental Protection Agency. "Wetlands Compensatory Mitigation." Accessed October 4, 2019. https://www.epa.gov/sites/production/files/2015-08/documents/compensatory_mitigation_factsheet.pdf.

U.S. Fish and Wildlife Service. "Bipartisan Infrastructure Law Funds Proven Projects for Wildlife." Accessed July 11, 2023. https://www.fws.gov/initiative/directors -priorities/bipartisan-infrastructure-law-funds-proven-projects-wildlife.

———. "Endangered Species Act: Grants: Overview." Accessed November 17, 2021. https://www.fws.gov/endangered/grants.

———. "Environmental Conservation Online System: Conservation Plans by Type and Region." Accessed August 7, 2023. https://ecos.fws.gov/ecp/report/conservation -plans-type-region.

———. "Environmental Conservation Online System: Listed Species Summary (Box-score)." Accessed December 13, 2023. https://ecos.fws.gov/ecp/report/boxscore.

———. "Environmental Conservation Online System: Red-Cockaded Woodpecker (*Picoides borealis*)." Accessed October 3, 2018. https://ecos.fws.gov/ecp0/profile/ speciesProfile?spcode=B04F.

———. "Environmental Conservation Online System: Species Reports: Delisted Species." Accessed December 13, 2023. https://ecos.fws.gov/ecp/report/species -delisted.

———. "Environmental Conservation Online System: USFWS Threatened & Endangered Species Active Critical Habitat Report." Accessed December 13, 2023. https: //ecos.fws.gov/ecp/report/critical-habitat.

———. "Monarch Conservation Database." Accessed November 12, 2021. https:// storymaps.arcgis.com/stories/8169fcb99632492cb73871654b9310bb.

———. "Sport Fish Restoration Apportionments." *Wildlife and Sport Fish Restoration Program*. Accessed March 7, 2023. https://tracs.fws.gov/sportFishRestorationAp portionments.html.

———. "State Wildlife Grant Program—2020 Apportionment." https://www.fws.gov /wsfrprograms/Subpages/GrantPrograms/SWG/SWG2020Apportionment.pdf.

———. "State Wildlife Grant Program—2021 Apportionment." https://www.fws.gov /wsfrprograms/Subpages/GrantPrograms/SWG/SWG2021Apportionment.pdf.

———. "State Wildlife Grant Program Apportionment History." 2019. https:// wsfrprograms.fws.gov/Subpages/GrantPrograms/SWG/SWGAppt2002-2019 .pdf.

———. "West Fork Timber HCP (Formerly Murray Pacific)." Accessed April 23, 2021. https://ecos.fws.gov/ecp0/conservationPlan/plan?plan_id=116.

———. "Who's on the Lek: A Guide to Players." Accessed August 11, 2017. https://www .fws.gov/mountain-prairie/species/birds/sagegrouse/Primer2SGWhoOnTheLek .pdf.

———. "Wildlife Restoration and Hunter Education Apportionments." *Wildlife and Sport Fish Restoration Program*. Accessed March 7, 2023. https://tracs.fws.gov/wild lifeRestorationAndHunterEducationApportionments.html.

———. "Wolf Livestock Loss Demonstration Project Grant Program." Accessed June 20, 2023. https://www.fws.gov/service/wolf-livestock-loss-demonstration-project -grant-program.

U.S. Forest Service. "Bat Pollination." Accessed November 11, 2021. https://www.fs.fed .us/wildflowers/pollinators/animals/bats.shtml.

———. "Bird Pollination." Accessed November 11, 2021. https://www.fs.fed.us/wildflowers/pollinators/animals/birds.shtml.

———. "Northwest Forest Plan Aquatic Conservation Strategy." Accessed June 3, 2021. https://www.fs.fed.us/r6/reo/acs.

———. "Northwest Forest Plan Overview." Accessed April 19, 2021. https://www.fs.fed.us/r6/reo/overview.php.

Watkins, Tate. "Endangered Frog's Survival Depends on Making Landowners Friends Not Foes." *Property and Environment Research Center.* September 27, 2018. https://www.perc.org/2018/09/27/endangered-frogs-survival-depends-on-making-landowners-friends-not-foes.

Wolf Conservation Center. "What Draws Tourists to Yellowstone? Wolves!" May 18, 2017. https://nywolf.org/2017/05/what-draws-tourists-to-yellowstone-wolves.

Woods, Josh. "US Beekeepers Continue to Report High Colony Loss Rates, No Clear Progression toward Improvement." *Auburn University.* Updated June 25, 2021, and June 24, 2021. https://ocm.auburn.edu/newsroom/news_articles/2021/06/241121-honey-bee-annual-loss-survey-results.php.

Working Together for the New England Cottontail. "Partners." Accessed January 28, 2021. https://newenglandcottontail.org/partners.

World Economic Forum. "3 Reasons Companies Are Investing in Forest Conservation and Restoration, and How They Do It." Accessed November 19, 2021. https://www.weforum.org/agenda/2021/06/3-reasons-companies-are-investing-in-forest-conservation-and-restoration-and-how-they-do-it.

World Wildlife Fund. "How Many Species Are We Losing?" Accessed October 27, 2021. https://wwf.panda.org/discover/our_focus/biodiversity/biodiversity.

———. "Living Planet Report 2020." Accessed October 28, 2021. https://livingplanet.panda.org/en-us.

Xerces Society. "Monarchs in Decline." Accessed September 18, 2023. https://xerces.org/monarchs/conservation-efforts.

———. "Western Monarch Conservation." Accessed July 19, 2023. https://xerces.org/monarchs/western-monarch-conservation.

———. "Who Are the Pollinators?" Accessed November 11, 2021. https://www.xerces.org/pollinator-conservation/about-pollinators.

Yale University Library. "Sustainability: Corporate Sustainability." Accessed November 19, 2021. https://guides.library.yale.edu/c.php?g=296179&p=2582471.

Zattara, Eduardo E. "Historic Data Hint at a Worldwide Decline of Wild Bee Biodiversity." *Indiana University Bloomington.* January 22, 2021. https://biology.indiana.edu/news-events/news/2021/wild-bee-decline.html.

OTHER SOURCES

Association of Fish and Wildlife Agencies. *Best Practices for State Wildlife Action Plans: Voluntary Guidance to States for Revision and Implementation.* November 2012. https://www.fishwildlife.org/application/files/3215/1856/0300/SWAP_Best_Practices_Report_Nov_2012.pdf.

———. *The State Conservation Machine*. 2017. http://www.fishwildlife.org/files/The
_State_Conservation_Machine-FINAL.pdf.

Bean, Michael J. *Endangered Species Act Safe Harbor Agreements: An Assessment Working
Paper*. Madison, WI: Sand County Foundation/Environmental Policy Innovation
Center, 2017.

———. *Habitat Exchanges: A New Tool to Engage Landowners in Conservation Working
Paper*. Madison, WI: Sand County Foundation/Environmental Policy Innovation
Center, 2017.

———. *Landowner Assurances under the Endangered Species Act Working Paper*. Madison,
WI: Sand County Foundation/Environmental Policy Innovation Center, 2017.

Center for Biological Diversity. *Petition to List 404 Aquatic, Riparian and Wetland Species
from the Southeastern United States as Threatened or Endangered under the Endangered Species Act*. April 20, 2010. https://www.fws.gov/southeast/pdf/petition/404
-aquatic.pdf.

Center for Biological Diversity and Humane Society of the United States. *Emergency
Petition to Relist Gray Wolves (*Canis lupus*) in the Northern Rocky Mountains as an
Endangered or Threatened "Distinct Population Segment" under the Endangered Species
Act*. May 26, 2021. https://ecos.fws.gov/docs/tess/petition/992.pdf.

Center for Biological Diversity, James D. Williams, and Zygmunt Plater. *Before the Secretary of the Interior: Petition to Delist the Snail Darter under the Endangered Species
Act*. July 16, 2019.

Commission for Environmental Cooperation. *North American Monarch Conservation
Plan*. 2008. https://www.fs.fed.us/wildflowers/pollinators/Monarch_Butterfly/
news/documents/Monarch-Monarca-Monarque.pdf.

Connelly, John W., Steven T. Knick, Michael A. Schroeder, and San J. Stiver. *Conservation Assessment of Greater Sage-Grouse and Sagebrush Habitats*. Boise, ID: Western
Association of Fish and Wildlife Agencies, 2004.

Defenders of Wildlife. *Southwest Wolf Compensation Guidelines*. https://defenders
.org/sites/default/files/publications/instructions_for_wolf_compensation_in_the
_southwest.pdf.

Fuller, Steven, Anthony Tur, and New England Cottontail Technical Committee. *Conservation Strategy for the New England Cottontail (Sylvilagus transitionalis)*. November 20, 2012. https://newenglandcottontail.org/sites/default/files/conservation
_strategy_final_12-3-12.pdf.

Humane Society of the United States. *Government Data Confirm That Wolves Have
a Negligible Effect on U.S. Cattle & Sheep Industries*. March 6, 2019. https://
www.humanesociety.org/sites/default/files/docs/HSUS-Wolf-Livestock-6.Mar_
.19Final.pdf.

Interagency Grizzly Bear Study Team. *Yellowstone Grizzly Bear Investigations 2022: Annual
Report of the Interagency Grizzly Bear Study Team*. 2023. https://igbconline.org
/wp-content/uploads/2023/09/Yellowstone-Grizzly-Bear-Investigations-2022
-IGBST-Annual-Report.pdf.

Johnson, K. Norman, Jerry F. Franklin, Jack Ward Thomas, and John Gordon. *Alternatives for Management of Late-Successional Forests of the Pacific Northwest: A Report to*

the Agricultural Committee and the Merchant Marine Committee of the U.S. House of Representatives. October 8, 1991.

Memorandum of Understanding among U.S. Fish and Wildlife Service, National Alliance of Forest Owners, and National Council for Air and Stream Improvement, Inc. 2023. https://nafoalliance.org/wp-content/uploads/2023/03/Wildlife-Conservation-Initiative-MOU-USFWS-NAFO-NCASI.pdf.

Memorandum of Understanding among Western Association of Fish and Wildlife Agencies and U.S. Department of Agriculture, Forest Service and U.S. Department of the Interior, Bureau of Land Management and U.S. Department of the Interior, Fish and Wildlife Service. 2008. https://www2.usgs.gov/mou/docs/mou_sage_grouse.pdf.

Midwest Association of Fish and Wildlife Agencies. *Mid-America Monarch Conservation Strategy 2018–2038.* http://www.mafwa.org/wp-content/uploads/2019/02/MAMCS_Summary_Handout_web.pdf.

———. *Mid-America Monarch Conservation Strategy.* 2018. http://www.mafwa.org/wp-content/uploads/2018/05/MidAmericaMonarchStrategyDraft_May11_2018.pdf.

Monarch Joint Venture. *2021 Monarch Conservation Implementation Plan.* 2021. https://monarchjointventure.org/images/uploads/documents/2021_Monarch_Conservation_Implementation_Plan.pdf.

National Fish and Wildlife Foundation. *2022 Conservation Investments.* https://www.nfwf.org/sites/default/files/2023-03/nfwf-2022-conservation-investments-online.pdf.

National Security Strategy. October 12, 2022. https://www.whitehouse.gov/wp-content/uploads/2022/10/Biden-Harris-Administrations-National-Security-Strategy-10.2022.pdf.

Oregon Department of Fish and Wildlife. *Greater Sage-Grouse Conservation Assessment and Strategy for Oregon: A Plan to Maintain and Enhance Populations and Habitat.* April 22, 2011. https://www.dfw.state.or.us/wildlife/sagegrouse/docs/20110422_GRSG_April_Final%2052511.pdf.

Petersen, Shannon. "The Modern Ark: A History of the Endangered Species Act." PhD dissertation, University of Wisconsin, Madison, 2000.

Range-Wide Interagency Sage-Grouse Conservation Team. *Near-Term Greater Sage-Grouse Conservation Action Plan.* September 11, 2012. http://sagemap.wr.usgs.gov/docs/rs/NTSGConservation%20Action%20Plan.pdf.

Regenstein, Lewis G. "A History of the Endangered Species Act of 1973 and an Analysis of Its History; Strengths and Weaknesses; Administration; and Probable Future Effectiveness." MA thesis, Emory University, 1975.

Remington, T. E., P. A. Deibert, S. E. Hanser, D. M. Davis, L. A. Robb, and J. L. Welty. *Sagebrush Conservation Strategy—Challenges to Sagebrush Conservation.* U.S. Geological Survey Open-File Report 2020-1125. 2021. https://doi.org/10.3133/ofr20201125.

San Bruno Mountain Habitat Conservation Plan Steering Committee. *San Bruno Mountain Area Habitat Conservation Plan—Final.* November, 1982.

Seasholes, Brian. *Fulfilling the Promise of the Endangered Species Act: The Case for an Endangered Species Reserve Program*. Reason Foundation. 2014. https://reason.org/wp-content/uploads/files/endangered_species_act_reform.pdf.

Secretary of State, State of Colorado. *2020 Abstract of Votes Cast*. 2020. https://www.sos.state.co.us/pubs/elections/Results/Abstract/2020/2020BiennialAbstractBooklet.pdf.

Stiver, S. J., A. D. Apa, J. R. Bohne, S. D. Bunnell, P. A. Deibert, S. C. Gardner, M. A. Hilliard, C. W. McCarthy, and M. A. Schroeder. *Greater Sage-Grouse Comprehensive Conservation Strategy*. Cheyenne, WY: Western Association of Fish and Wildlife Agencies, 2006. https://wdfw.wa.gov/sites/default/files/publications/01317/wdfw01317.pdf.

Talbot, Lee. "Patience and Tenacity: The Endangered Species Act of 1973" (unpublished document).

"A Timeline of Sage Grouse Conservation." *Western Confluence*. 2013. http://www.sagegrouseinitiative.com/wp-content/uploads/2014/03/Western-%0AConfluence-»-A-timeline-of-sage-grouse-conservation.pdf.

Western Association of Fish and Widlife Agencies. "Western Monarch Butterfly Conservation Plan, 2019–2069." Accessed July 19, 2023. https://wafwa.org/wpdm-package/western-monarch-butterfly-conservation-plan-2019-2069.

Western Watersheds Project et al. *A Petition to List the Western North American Population of Gray Wolves (Canis lupus) as a Distinct Population Segment*. July 29, 2021. https://ecos.fws.gov/docs/tess/petition/3352.pdf.

Wikipedia.

Williams, John, Douglas E. Johnson, Patrick E. Clark, Larry L. Larson, and Tyanne J. Roland. "Wolves—A Primer for Ranchers, EM 9142." Oregon State University, March 2017. https://catalog.extension.oregonstate.edu/em9142/html.

Wilson, Seth M. *Summary of Grizzly and Wolf Conflicts for Blackfoot Watershed: Report to Montana Livestock Loss Board*. February 6, 2021.

World Wildlife Fund. *Living Planet Report 2020: Bending the Curve of Biodiversity Loss*. 2020. https://f.hubspotusercontent20.net/hubfs/4783129/LPR/PDFs/ENGLISH-FULL.pdf.

Wright, Katherine, and Shawn Regan. "Missing the Mark: How the Endangered Species Act Falls Short of Its Own Recovery Goals." *Property and Environment Research Center*. July 26, 2023. https://www.perc.org/2023/07/26/missing-the-mark.

Wyo. H. J. No. 13. Resolution in Support of Central Park Wilderness. 2012.

INDEX

ABOUT THE AUTHOR

Author photo by Len Spoden

Lowell E. Baier received his BA in economics and political science from Valparaiso University in 1961 and completed his law degree in 1964 at the Indiana University School of Law, where he earned a JD. After graduation, he practiced law in Washington, D.C., where he has devoted his career to his lifelong passion for protecting the country's natural resources and wildlife conservation. Baier holds an honorary doctor of law and letters degree awarded in 2010, a doctor of humane letters (LHD) awarded in 2015, and a doctor of public service awarded in 2019. From 2018 to 2020, Baier was enrolled at the University of St. Andrews, Scotland, and completed coursework and his dissertation for a doctor of philosophy in environmental history. Thereafter, he published his dissertation titled *Federalism, Preemption, and the Nationalization of American Wildlife Management: The Dynamic Balance between State and Federal Authority*.

Baier has been recognized many times for his extraordinary public service at the local level and for his conservation work nationally. Since 1975, Baier has been active in the Boone and Crockett Club, America's oldest wildlife conservation organization founded by Theodore Roosevelt in 1887, and is its first president emeritus. A well-known adviser to elected officials and educators on environmental and conservation issues,

Baier took the lead in drafting President George H. W. Bush's wildlife conservation agenda in 1989 and has been an adviser and counselor to all successive presidential administrations. He was Citizen of the Year for Rockville, Maryland, in 1986. In 2008, he was named Conservationist of the Year by the National Fish and Wildlife Foundation. In 2010, *Outdoor Life* magazine selected Baier as the Conservationist of the Year, and the Association of Fish and Wildlife Agencies similarly recognized him in 2013. In 2016, the National Wildlife Federation awarded him the organization's highest honor, the Jay N. "Ding" Darling Conservation Award for a lifetime of conservation service. In 2018, he was one of four judges chosen to select the 2019 Duck Stamp image by the Department of the Interior's U.S. Fish and Wildlife Service.

The Indiana University School of Law presented him with its Distinguished Service Award in 2007, and in 2014 he was inducted into the Academy of Law Alumni Fellows, the highest honor the school can bestow on an alumnus. In 2015, Indiana University awarded him the LHD degree; thereafter, also in 2015, the law school building at Indiana University was named Baier Hall in his honor. In 2021, he was awarded the university's Bicentennial Medal.